T0138572

International Journal of Space Studies is a joint venture between ATF Press Publishing Group, the International Space University and the University of South Australia. The aim of the Journal is to bring together research undertaken by students in the programs coordinated by ISU. Each edition has its own theme and editorial team who work under the supervision of the overall editors from both educational institutions with an International Reference group.

Space Assets and Technologies for Bushfire Management

Edited by
Travis Holland and Zoë Nicole Silverstone

The 2021 Southern Hemisphere Space Studies Program was held by the International Space University (ISU) and the University of South Australia (UniSA).

**University of
South Australia**

University of South Australia
Mawson Lakes Boulevard
Mawson Lakes
South Australia 5095
www.unisa.edu.au
Electronic copies of the Executive Summary and Team Project Report may be found on the
ISU website (https://isulibrary.isunet.edu/).

International Space University
Strasbourg Central Campus Parc d'Innovation
1 rue Jean-Dominique Cassini 67400 Illkirch-Graffenstaden France
Tel +33 (0)3 88 65 54 30
Fax +33 (0)3 88 65 54 47
e-mail: publications@isunet.edu
website: www.isunet.edu

The cover depicts the GOES-16 satellite orbiting Earth with Australian Wildfires reflecting in its solar array panels.
Images courtesy of SHSSP21 Graphics Team. Photo of GOES-16 satellite courtesy of NASA.
Australian Wildfires captured by Sentinel-2, image courtesy of The European Space Agency.

While all care has been taken in the preparation of this Team Project report, ISU does not take any responsibility for the accuracy of its content.

Space Assets and Technologies for Bushfire Management

Edited by
Travis Holland and Zoë Nicole Silverstone

Space
Adelaide
2021

© International Space University. All Rights Reserved

Editors: Travis Holland and Zoë Nicole Silverstone

Copy editor: Gabriel Bueno Siqueira

Cover design: Cristian Gonzalez Guerrero, Myf Cadwalladerl, Nyssa Lonsdale, Zandria Farrell

Layout: Extel Solutions, India

Font: Mino Pro

Title: International Journal of Space Studies, Volume 1, 2021

Managing Editor: Gabriel Bueno Siqueira, ATF Press Adelaide

Editorial Board Representing University of South Australia: Ady James, Jasmine Vreugdenber.

Representing the International Space University: Carol Carnett, Kenos Jules.

International Board of Reference: Jacques Arnould (CNES France), Michael Davis, (Andy Thomas Space Foundation, Adelaide), Yi So-yeon (ISU).

ISBN: 978-1-922582-81-2 soft cover
               978-1-922582-82-9 hard cover
               978-1-922582-83-6 epub
               978-1-922582-84-3 pdf

Published by

Space

Expanding Horizons

An imprint of the ATF Press Publishing Group
owned by ATF (Australia) Ltd.
PO Box 234
Brompton, SA 5007
Australia
ABN 90 116 359 963
www.atfpress.com

International Journal of Space Studies, Volume 1, 2021

Table of Contents

Acknowledgements

The International Space University Southern Hemisphere Space Studies Program 2021 and the work on the team project were made possible through the generous support of the following organizations:

Program Sponsors

ATF Press

Lockheed Martin

Ten to the Ninth plus Foundation

Scholarship Providers

European Space Agency (ESA)

Government of South Australia

Italian Space Agency (ISA)

Sir Ross and Keith Smith Fund

Stichting Space Professionals Foundation (SSPF)

Sponsored Placements

Australian Space Agency (ASA)

CNES

IBM

Italian Air Force

KPMG Australia

LafargeHolcim

Philippines Space Agency

Saudi Space Commission

SmartSat CRC

Event Sponsors

Sir Ross and Keith Smith Fund

The authors gratefully acknowledge the generous guidance, support, and direction provided by the following faculty, visiting lecturers, program officers, program staff, advisors, and experts:

Staff	Instructors
Adrian James	Adrian James
Alexandra Ryan	Alev Sonmez
Amanda Johnston	Alice Gorman
Amanda Michelle Simran Sathiaraj	Allison Kealy
Anisha Rajmane	Anna Maria Kleregard
Ankita Das	Bernard Foing
Arif Goktug Karacalioglu	Brett Gooden
Brenton Dansie	Caley Burke
Camilo Reyes	Carol Carnett
Carol Carnett	Charley Lineweaver
David Cowdrey	Charlotte Pouwels
David Harris	Christopher Welch
Eric Dahlstrom	Danijela Stupar
Geraldine Moser	Edward Michael Fincke
Hannah Webber	Federico Rondoni
Jack Fraser	Gary Martin
Joel Herrmann	George Dyke
Jonas Tobiassen	Giuliana Rotola
Juan Carlos Mariscal	Iliass Tanouti
Kavindi De Silva	James Green
Kenol Jules	John Connolly
Konstantin Chterev	Joseph Ibeh
Maïssa El Moussaoui	Juan de Dalmau
Mirela Souza De Abreu	Kate Sweatman
Muriel Riester	Kerrie Dougherty
Nathan Taylor	Kimberley Norris
Nicolas Moncussi	Laura ten Bloemendal
Ryan Clement	Léo Baud
Sasiluck Thammasit	Liad Yosef
Sebastien Bessat	Loredana-Beatrice Teodor
Supprabha Nambiar	Marc Heemskerk
Supreet Kau	

Mentors

Abhishek Akash Diggewadi
Alberto De Paula Silva
Alejandro Ignacio Lopez Telgie
Ana Cristina Galhego Rosa
Anis Karim
Anna Maria Kleregard
Antonio Fortunato
Caley Burke
Carol Carnett
Cenan Al-Ekabi
Christian Thaler-Wolski
Claudiu Mihai Taiatu
Cristina Cerioni
Daniel Rockberger
Danna Linn Barnett
David Bruce
David Hyland-Wood
Elena Evgenjevna Grashchenkova
Filip Novoselnik
Garrett Smith
George Dyke
Jason Hair
Jeremy Myers
John Connolly
Jonathan Grzymisch
Jose Ocasio-Christian
Juha-Pekka Luntama
Kacper Grzesiak
Kate Sweatman
Ke Wang
Krzysztof Kanawka

Liad Yosef
Lolowa Alkindi
Lucy Stojak
Manuel Antonio Cuba
Marc Heemskerk
Marco Marsh
Maurizio Nati
Michael Davis
Michael François
Narashima Purohit Paranjothy Karunaharan
Paul Iliffe
Perry Edmundson
Piero Messina
Rahul Goel
Robert Hunt
Rowena Christiansen
Sabrina Alam
Samy Nicolas Bouchalat
Scott Ritter
Sean Madden
Shrrirup Nambiar
Steven Brody
Susan Ip Jewell
Su-Yin Tan
Tommaso Tonina
Uri Greisman Ran
Vienna Tran
Zuo Zheng

Coaches

Adrian James
Carol Carnett
Dimitra Stefoudi
Gary Martin
Lucy Stojak
Madhu Thangavelu
Michael Simpson
Niamh Shaw
Ray Williamson
Ruth McAvinia
Tanja Masson-Zwaan

TP project matter experts

Cormac Purcell
Ian Tanner
Mike Wouters
Shaye Hatty
Craig Leach
Laura Ten Bloemendal
Helena van Mirelo
Loredana-Beatrice Teodor

List of Participants

The class of SHSSP21 comprised and was enriched by the following Participant Members from around the globe:

Australia

Michael Burnside	Rebecca Kuster	Ryan Roberts
Oliwia Derda	Nyssa Lonsdale	Caolan Rohleder
Zandria Farrell	Samantha Page	Nipuni Silva
Isabella Hatty	Toby Rady	Zoë Silverstone
Nicola Higgins	Samantha Raines	Dharshun Sridharan
Travis Holland	Siddharth Rajput	Emily White

Ecuador

Robert Aillon

Eritrea

Helen Haile

Philippines

Jamaica Ida Palce

Saudi Arabia

Yara Aljarallah
Safwan Najjar

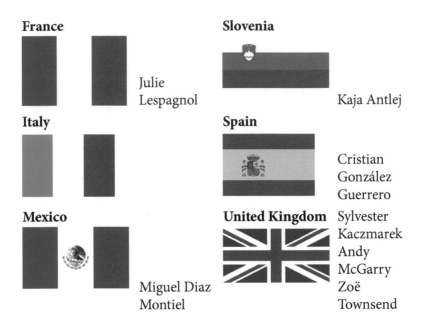

France

Julie
Lespagnol

Italy

Mexico

Miguel Diaz
Montiel

Slovenia

Kaja Antlej

Spain

Cristian
González
Guerrero

United Kingdom

Sylvester
Kaczmarek
Andy
McGarry
Zoë
Townsend

International Journal of Space Studies, Volume 1, 2021

Abstract

The financial, emotional, and ecological impacts of bushfires can be devastating, as seen during the 2019–2020 'Black Summer' East Australian bushfires. This report was prepared by participants of the Southern Hemisphere Space Studies Program 2021 (SHSSP21) to address the question, *'How can space assets and technologies be applied to better predict and mitigate bushfires and their impacts?'* The task was addressed under three pillars, namely bushfire prediction ('Predict'), mitigation of bushfire effects ('Mitigate'), and communication needs relating to bushfire response ('Communicate'). Working initially in these three groups, the report was developed by combining the diverse expertise and experience of participants with the interdisciplinary knowledge gained from seminars, distinguished lectures, and workshops, and a literature review. With reference to the 2019–2020 Australian fire season, each group conducted a current state analysis and identified key challenges. Comparing this to the future desired state, we identified gaps in each of the three domains, and then worked across teams to reach consensus on a combined list of recommendations. Several recommendations were derived independently by two or more of the three groups, highlighting the importance of a holistic and collaborative approach. In summary, three geophysical components determine fire behavior—fuel, topography and weather. Earth observation data for each of these are essential for accurate bushfire prediction. As most fires start by lightning strikes, prediction of lightning using satellite infrared imagery to determine cloud top temperatures could assist bushfire prediction. Huge volumes of satellite data must be combined with weather and climate prediction models to improve forecast lead times, however, a major limitation is the storage capacity and computational power required, which could be addressed by using cloud-based technology infrastructure. Adequate communication of relevant information to authorities and communities is vital. Firefighter safety could be improved by use of protective clothing modelled on astronaut space suits, and by reinforcing fire trucks with thermal protective materials as used on spacecraft. An interoperable national communications infrastructure is needed to enable rapid sharing of information and resources, with emergency responders and citizens, across jurisdictions. While our case study was based on the Australian context, the lessons and recommendations could be applied internationally.

International Journal of Space Studies, Volume 1, 2021

Faculty Preface

In recent years, bushfires of unprecedented scope have devastated Australia, the western United States of America (USA), Brazil, Siberia, and other locations around the world. While firefighters and governments work to deal with the scale of these megafires, new tools are becoming available through the use of satellite systems and other space technologies. This team project examined how space assets and technologies could be applied to better predict and mitigate bushfires and their impacts.

The Southern Hemisphere Space Studies Program of the International Space University is an intensive five-week program covering all aspects of space. This year the program was operated virtually and conducted over Australian and European time zones. For the bushfire topic, the participants focused on Australian bushfires but within an international context as the world deals with a changing climate.

The team project is an important culmination of the program requiring participants to integrate their interdisciplinary lessons and apply them to a specific topic. This year, thirty-three participants from eleven countries, with expertise ranging from engineering, law, computer science, management, marketing, education, meteorology, to medicine, all combined their efforts to examine space assets applied to bushfires. The participants organized themselves into teams that spanned time zones around the world to examine the many interdisciplinary aspects of space assets and technologies for bushfire management.

I wish to thank the global team that supported this program, and I congratulate the 33 authors for writing this important report.

International Journal of Space Studies, Volume 1, 2021

Eric Dahlstrom, Team Project Chair

Participants' Preface

The 2021 Southern Hemisphere Space Studies Program (SHSSP21) run by the International Space University (ISU) and the University of South Australia (UniSA) included thirty-three students from around the world. Due to the coronavirus 2019 (COVID-19) pandemic, the program was conducted in an online format over a five-week period from January to February. Participants ranged from twenty to fifty-six years of age, and came from diverse cultures and backgrounds, with past experience ranging from arts, science, and technology to humanities. Importantly, fifty-five per cent of participants were women. Lectures, workshops and project sessions ran across two time-zones (broadly, 'Australian' and 'European') with joint sessions at the juncture of the day. We all had to adapt to an intense but amazing program of space-related lectures and workshops conducted in an online video conferencing forum. We benefitted from highly esteemed international experts who generously shared their time and their wisdom.

A requirement of the program was that participants conduct a team project with written report and presentation. This report represents the written component of this task. The team project topic chosen by ISU and UniSA for SHSSP21 is particularly relevant to this class, many of whom are from Australia and experienced the effects and disruption of that catastrophic fire season over the 2019-2020 summer. Some participants have connections to various Australian fire and emergency services. Although the Australian fires loomed large, we know that wildfires are becoming an increasingly prominent global problem, exacerbated by climate change. In recent years, large fires burnt through tracts of land at high latitudes in Siberia and Canada as well as in the United States of America (USA), parts of Europe, Asia, and in the Amazon in South America. The class was therefore highly motivated to research the problem and to make recommendations. The hope is that this fresh perspective, built within the ISU's interdisciplinary, international, and intercultural approach, could help in some way to tackle this natural hazard in Australia and around the world.

For those in the class not from Australia, it has been a wake-up call to hear first-hand about the devastation and loss of life and property, the destruction of the environment, and the impact on wildlife. One of the most poignant moments was when we were able to interview, via video call, the relatives of one of the participants and hear directly about what it means to be a firefighter in Australia's Northern Territory. We also interviewed experienced firefighters from South Australia's Country Fire Service and it was remarkable to hear how they welcomed our efforts to help them, and how interested they were in the project. We were inspired by their enthusiasm to see the results and recommendations arising from our work.

Remote working technology, including video conferencing software and a collaborative shared online workspace, has enabled us to work across time zones and hemispheres. We were able to collaborate effectively and even work almost around the clock. For all of us this has been an incredible learning experience that has forged new bonds and friendships around the world in a short space of time.

None of us will forget this incredible experience, and we are immensely thankful to the staff, tutors, coaches, mentors, sponsors and subject-matter experts who made this possible.

Table of Figures

International Journal of Space Studies, Volume 1, 2021

List of Tables

List of Acronyms

2D	Two-dimensional
3D	Three-dimensional
5G	Fifth generation telecommunications standard, for cellular networks
AFAC	Australasian Fire and Emergency Services Authorities Council
AFDRS	Australian Fire Danger Rating System
AGCMF	Australian Federal Government Crisis Management Framework
AI	Artificial Intelligence
ANU	Australian National University
AR	Augmented Reality
AWS	Australian Warning System
BNHCRC	Bushfire and Natural Hazards CRC
BoM	Bureau of Meteorology
CAVE	Cave Automatic Virtual Environment
CRC	Cooperative Research Centre
CSIRO	Commonwealth Scientific and Industrial Research Organization
Data61	Australia's data innovation network, run by CSIRO
EarthCARE	Earth Cloud, Aerosol and Radiation Explorer mission
ENSO	El Niño-Southern Oscillation

EO	Earth Observation
FLAMA	Fire Logistics and Management Approach
GA	Geoscience Australia
GEO	Geostationary Orbit
GIS	Geographic Information System
GLM	Geostationary Lightning Mapper instrument
GNSS	Global Navigation Satellite System
GOES	Geostationary Operational Environmental Satellite
GPATS	Global Position and Tracking Systems
HMD	Head Mounted Display
IR	Infrared
ISU	International Space University
JMA	Japan Meteorological Agency
LEO	Low Earth Orbit
LiDAR	Light Detection and Ranging (a remote sensing method)
MODIS	Moderate Resolution Imaging Spectroradiometer
NASA	National Aeronautics and Space Administration
NSSL	National Severe Storms Laboratory
SAR	Synthetic Aperture Radar
SHSSP21	Southern Hemisphere Space Studies Program 2021
SMS	Short Message Service
UAV	Unmanned Aerial Vehicles
UHF	Ultra High Frequency
UN	United Nations
UniSA	University of South Australia
USA	United States of America
VR	Virtual Reality
XR	Extended Reality

1
Introduction

1.1 Background

The 2019–2020 East Australian bushfires were the most severe on record. Tragically, there were thirty-three deaths directly due to the fires, many of whom were firefighters (Commonwealth of Australia, 2020). There were also an estimated 400 excess deaths and over 3000 excess hospitalizations due to smoke related air-pollution (Borchers-Arriagada *et al*, 2020). Three thousand homes were destroyed and there was an estimated \$10 billion in financial losses. Many communities became trapped and isolated, and failure of critical infrastructure such as power and telecommunications left many without access to essential services and payment facilities for prolonged periods of time. The 2020 Royal Commission into National Natural Disaster Arrangements (Royal Commission), while identifying many strengths in the emergency response, also identified a number of shortcomings. The conclusion was that we need to be better prepared for such events in the future, particularly given they are likely to become more frequent because of climate change (Commonwealth of Australia, 2020).

Australia is not alone in the battle against bushfires. Catastrophic fires are an increasingly common occurrence around the world, including in the United States of America (USA) (2018, 2020), Portugal (2017), Greece (2018), Brazil (2019) and Colombia (2019). The traumatic impact on those dealing with these events, including front-line firefighters and affected communities, are long-lasting. Firefighters, like astronauts, face challenges in dealing with extreme conditions. This requires effective management, planning and interdisciplinary approaches where people, process, and technology together affect the response and outcomes.

Space assets and technologies have already made wide-reaching contributions to human society, in areas including aviation, global mapping and navigation, communication, human health and life support, agriculture and food security, environment and climate monitoring, and disaster management (United States Office for Outer Space Affairs, 2012). In this report, we specifically explore how space assets and technologies can be used for the purposes of fighting future bushfires.

1.2   Report Purpose

We aimed to extend the findings of the Royal Commission, by identifying how space assets and technologies can be applied to better predict and mitigate bushfires and their impacts. We initially undertook an analysis of the current state of bushfire response and space asset usage, identified gaps and areas where existing systems could be improved, and used this analysis to form recommendations. Where applicable, we have aligned our discussion with the international context. We did not explicitly consider the detailed financial or economic implications of our recommendations.

1.3   Methodology and Approach

The team project was conducted as a component of the ISU SHSSP21. We initially identified three pillars of an effective fire response, namely, accurate and timely prediction of bushfires (*'predict'*), effective mitigation of bushfire impacts (*'mitigate'*), and timely communication of relevant information to all key stakeholders (*'communicate'*). We worked in three groups to explore each of these aspects. We performed several rounds of research, including literature review, in conjunction with a series of workshops and lectures from eminent international subject matter experts, and ISU staff. Once gaps had been identified for each of the three domains, we worked across teams to reach consensus on a list of recommendations. Several of these recommendations were derived independently by two or more of the three groups, highlighting the importance of a holistic and collaborative approach.

2
Current State Findings

The Royal Commission made a number of recommendations to enhance Australia's ability to respond to bushfires (Commonwealth of Australia, 2020). To assess the current state in relation to use of space technology, our groups reviewed these recommendations as they applied to the three pillars of this report, investigated changes that have been implemented since that review, conducted a review of other literature, and spoke with several subject matter experts. We focused specifically on aspects of bushfire prevention and impact mitigation that are, or could be, aided by space technologies.

2.1 Current State: Prediction of bushfires

Bushfires may travel over large distances at an incredibly rapid speed. Early detection, or even better, prediction of fires before they occur, can significantly reduce the extent of damage, difficulty, and the cost of the response. While it is impossible to predict all ignition events for all sources of fire, ability to predict the likelihood, severity and spread of fires would offer opportunities for advanced planning, asset placement, and community engagement in areas at risk.

Space assets and technologies may allow prediction of certain fires over timescales ranging from days, weeks, or months in advance of a fire season or fire event. We chose up to twelve months prior to ignition as the scope boundary for prediction. This long period prior to ignition offers the opportunity to predict fire potential across broad geographic areas, which would enable authorities and communities to adequately prepare. Prediction of ignition at specific locations could only be made over shorter timeframes.

Earth Observation (EO) data from multiple sources are essential for accurate prediction. We looked at international sources of data and space technologies that could benefit the prediction of fire specifically in Australia. The findings could be applied internationally.

The interplay of key factors relevant to fire prediction, including the scope-boundary between prediction and mitigation, is summarized in Appendix A.

Current fire prediction capabilities using space assets range from a few days to months, with analysis of fuel load moisture and other indicators conducted across seasons. Accurate weather predictions such as emerging lightning storms can help to pinpoint the location fires might occur with higher accuracy, but this is on a shorter timeframe than may be required for preparatory activities, including potential evacuations.

The use of space-based assets, such as satellites, to collect EO data has contributed to the prediction and management of bushfires, especially with the increased capability of remote-sensing technologies. There are several satellites with sensors that can detect land cover, fuel load, fuel moisture content and weather (Jain et al., 2020) (Table 1; Appendix B, C). These sensors routinely monitor vegetation distributions and changes, however, there are limitations. For example, detection of fuel moisture content by the Moderate Resolution Imaging Spectroradiometer (MODIS) tends to be good for European vegetation but is not as effective for Australian species (Yebra *et al*, 2018).

Satellites provide a plethora of data and remote sensing capabilities in Australia. Their use is rapidly expanding as a result of state government initiatives (Marshall, 2021), university-based projects (Australian National University, 2020), and new space start-ups (University of Southern Queensland, 2021). Combining data relating to land cover, fuel load and fuel moisture content with weather satellite prediction and climate models offers improved forecast lead times (Bauer *et al*, 2015).

Table 1 Representative satellites used in fire prediction applications

Satellite	LEO	GEO	In-service?	Relevance to bushfire prediction
EarthCARE	✓		No (end 2021)	Topography mapping, fuel load detection (including through cloud, smoke).
MODIS (Aqua Terra)	✓		Yes	Detection of fuel load and moisture content, land surface temperature and reflectance.
GOES-16		✓	Yes	Effective lightning detection (western hemisphere only)
Sentinel 2/3	✓		Yes	Fuel load and moisture content, topography measurements.

MODIS: Moderate Resolution Imaging Spectroradiometer; GOES: Geostationary Operational Environmental Satellite; EarthCARE: Earth Cloud, Aerosol and Radiation Explorer mission

2.1.1 Climate, weather, and lightning prediction

Gathering data on multiple aspects of weather conditions can contribute to long-term forecasts for fire prediction (Appendix D). Long-term forecasting is not an exact science, but provides critical insight into upcoming changing conditions.

Climate change plays a key role in the occurrence of bushfires. The *State of the Climate 2020* report noted: 'There has been an increase in extreme fire weather, and in the length of the fire season, across large parts of Australia since the 1950s' (BoM, 2020).

Fires start through lightning strikes (eighty-two per cent), accidents (fourteen per cent), arson (less than four per cent) or hazard reduction burns (one per cent) (Cook, 2020). There is no way to predict accident or arson but we can attempt to predict lightning strikes, severe weather events and storms that are likely to include electrical activity.

Lightning types include cloud-to-ground, cloud-to-cloud, or intra-cloud. Cloud-to-ground is the main concern for bushfire ignition, but the other types may precede cloud-to-ground lightning by five to ten minutes (NOAA, 2017) giving earlier warning for possible bushfires. While ground assets can detect lightning, including intra-cloud, space assets can detect it earlier, and over larger areas (NOAA, 2018). An instrument that specifically detects cloud-to-cloud and intra-cloud lightning is not currently available for Australia.

Infrared (IR) imagery can determine cloud top temperatures that may be used to predict the probability or intensity of lightning. Temperatures of around –11°Celsius may spark thunderstorms (Mäkelä, 2006). Satellite imagery can also provide data on troughs, sea breezes and gust fronts, which may trigger thunderstorms. These data allow meteorologists to predict where storms may develop. The greatest limitation is the computing power required to produce high resolution modelling.

Appendix D provides further detail on weather forecasting, climate change, lightning detection, and their relationship with bushfire prediction.

2.1.2 Fuel load and moisture content

The Commonwealth Scientific and Industrial Research Organization (CSIRO) noted that of the three components that combine to determine fire behavior—fuel, topography and weather—fuel is the only one that can be modified (Parliament of Australia, 2010). Vegetation condition also affects fire risk, contributing to heatwaves via the land-atmosphere feedback system (Tian *et al*, citing Seneviratne, 2019). Reducing the fuel hazard will reduce the overall danger posed by bushfires and increase the potential that a fire may be stopped through natural or artificial means (Parliament of Australia, 2010).

Space assets and technologies can offer substantial benefit in assessing fuel load and its moisture content. There is also potential to use root-zone soil water availability from satellite data to forecast vegetation condition at large scale (Tian *et al*, 2019).

Fuel moisture content changes over time, primarily due to weather, therefore recent observations are critical for accurate estimates. Satellite remote sensing has helped deepen our understanding of water availability and climate change, and has advanced predictive models and decision making (Tian et al. citing Rodell, 2019). Further work is required to reduce moisture content and flammability estimation error, and increase accuracy of risk assessment (Yebra *et al*, 2018).

Pre-emptive hazard reduction burning is one approach firefighters use to try to mitigate the spread and the effects of fires on human settlements. Some fires are predictable, and historical data has been captured to locate these areas; however, with a changing climate and more geographically distributed population, new at-risk areas need to be identified and catalogued.

Light Detection and Ranging (LiDAR) uses light in the form of a pulsed laser to measure distances to the Earth. The pulses of light are either reflected off a surface, or absorbed or scattered by particles in the atmosphere. The returning light pulses are detected in the scanner and the time it takes between sending and receiving is recorded. This is an example of an active scanner. Since LiDAR is a three-dimensional (3D) scanner it can reveal the topography of an area. When a fire burns uphill, the heat rises, causing the fire to spread faster. Use of LiDAR to identify hills with large fuel loads can highlight priority sites for hazard reduction burns.

Different sensor detection wavelengths across the electromagnetic spectrum (the spectrum) have varying strengths and weaknesses for the detection of fuel load and moisture content in natural landscapes. Detection also varies between active and passive sensors, so an integrated approach is required. Levels of integration maturity across sensor-types are varied, but progress is well underway.

2.1.3 Data integration maturity

Segregated EO data provides insight for only a subset of the whole fire-risk picture. This is evident in current data used for weather, climate, and disaster prediction, which does not allow an integrated comprehensive bushfire management strategy. Higher quality and more frequent input of EO data from multiple data sources will improve model accuracy.

The Bushfire Earth Observation Taskforce (Australian Space Agency, 2020) recommended that nationally-agreed data on infrastructure, such as powerlines, roads, trails, and telecommunications, were needed to support decision-making. LiDAR and radar satellite technology is known to have excellent capabilities to produce such data; however, the availability, resolution, and currency were cited as limitations.

Satellites are important for bushfire prediction, however, we rely on capacity for storage of satellite data. Space-deployed infrastructure with radar, spectral, and imagery systems generates corresponding

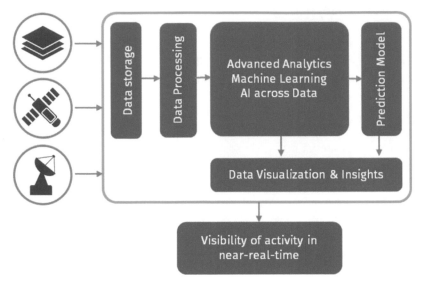

Figure 1: Prediction Data Platform

data. The data can then be processed through computational models to identify priority risk areas across the country, allowing appropriate deployment of hazard reduction methodologies (Figure 1).

2.2   Current State: Mitigation of bushfire impacts

2.2.1   Stakeholders and the fire management policy environment in Australia

Bushfire mitigation ultimately relies on changing human behavior. All levels of government in Australia have some role in mitigation, with most responsibility sitting with state government fire response agencies, local government authorities, and most importantly, the community directly at risk (CSIRO, 2020). Mitigation policies at the state and local government level are strong, but there are some inconsistencies.

The current annual awareness-raising campaigns by state governments, in collaboration with fire authorities, are designed to encourage the public to act with vigilance and implement individualized mitigation and survival plans. However, an Australian community-based study in 2010 revealed that only forty-three per cent of participants had prepared a personalized bushfire action plan and would likely delay or refuse to evacuate under bushfire threat, thereby compromising safety (Eriksen and Gill, 2010). It is clear that despite these various government outreach initiatives and campaigns, public awareness and threat preparedness remain low (Commonwealth of Australia, 2020).

Additionally, the Australian Space Agency (ASA) recognizes that stakeholder collaboration is essential for bushfire management. The Bushfire Earth Observation Taskforce, led primarily by ASA, hosted a series of targeted workshops with key stakeholders, organizations, and agencies in bushfire prevention and management to understand how they provide support (Australian Space Agency, 2020). Each state and territory government has slightly different bushfire management arrangements, which integrate and implement national frameworks and strategies differently (Commonwealth of Australia, 2020) (Figure 2).

In 2018, the National Disaster Risk Reduction Framework was created to establish priorities to reduce disaster risk in Australia. The framework recommends establishing a national coordination mechanism to monitor disaster risk reduction efforts across the country (Australian Government, 2018). The adoption of new space-based bushfire mitigation technologies will require support from a range of stakeholders at all levels, including the Federal Government. Targeted workshops would allow stakeholders to discuss the merits of different national frameworks and debate how best to share responsibilities between governments, and citizens, in the management of national disasters such as bushfires (McLennan and Eburn, 2012).

2.2.2 Technology

Firefighter safety

Frontline firefighting carries significant personal risk (Commonwealth of Australia, 2020). Environmental and other physical conditions are difficult to predict in real-time. These challenges are often

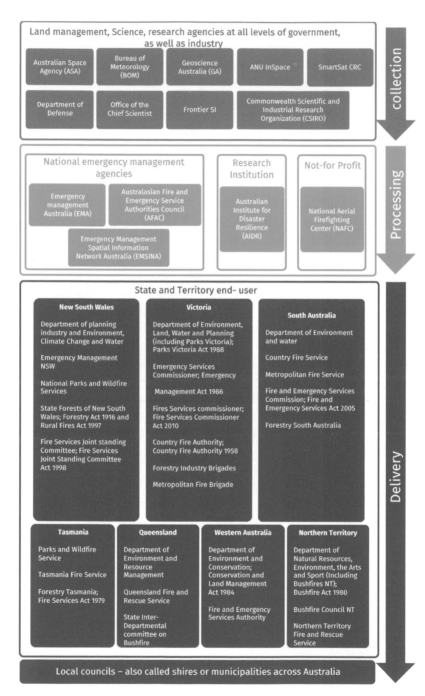

Figure 2: Key stakeholders in bushfire management in Australia

exacerbated by visibility constraints, delay in receiving time-critical data due to loss of communication, fatigue, psychological stress, oxygen deprivation, dehydration, and other stresses (Bush, 2015). It was noted during the 2019–2020 fires that these factors are often not considered in real-time until it is too late (CSIRO, 2020).

Fire-resistant personal protective equipment (PPE) is tested to withstand harsh conditions under nominal firefighting operations (NSW Fire & Rescue, 2021), however, during extreme events, particularly in the case of impaired cognitive responses, fatigue or if firefighters become trapped amongst flames, regular PPE may not offer optimal protection. Current burn-over protection technologies for firefighting vehicles are 'effective in less than fifty percent of historical burn overs resulting in firefighter injury or fatality' (Penney *et al*, 2020).

Sheltering facilities

There are variations in the definitions and specifications for fire sheltering facilities across states in Australia. Terminology and specifications need to be reviewed for consistency and clarity, guidelines should be prepared outlining the services and essential supplies to be provided at each sheltering facility (Commonwealth of Australia, 2020). As it stands, the average person may not know what support, services and level of safety are available at local shelters (Schauble, 2013).

2.3 Current State: Communication relating to bushfire response

2.3.1 Bushfire Management Responsibilities

Australia's bushfire and land management capabilities and policies are developed through a collaborative model, across agencies and emergencies, at all levels of government. Within this model the Australasian Fire and Emergency Service Authorities Council (AFAC) acts a lynchpin (Australian Space Agency, 2020).

We have identified an idealized hierarchy of key stakeholders for bushfire management in Australia (Figure 2). It is these government agencies and institutions that will manage the collection and processing of environmental data, and then delivery of information to end-users.

Although there is significant collaboration, there are distinct differences in the responsibilities and priorities of organizations operating at federal versus state and territory levels.

Fire response and land management agencies fall under the jurisdiction of the state and territory governments. As such they are charged with responsibility for the protection of life, property and the environment. The federal government provides support for these activities through the provision of emergency management and coordination of support to state and territory agencies, education and training, research and information sharing, scientific and technical assistance, and public awareness (Australian Senate, 2010).

The call for a national bushfire and land management policy, as well as a federal department to implement it, has been a common recommendation since Victoria's Black Saturday bushfires. Following a 2010 Senate inquiry, it was the committee's view that:

> It is not practical for the Commonwealth to have a more direct role in improving co-operation between fire agencies during bushfire emergencies, either within states or across state boundaries. The Commonwealth can assist with coordinating resources and developing standard operational systems across jurisdictions, but it is not for the Commonwealth to dictate to individual fire agencies the manner in which they co-operate with their intrastate counterparts or those across state borders (Australian Senate, 2010).

In recent years, funding allocated to fire management agencies has increased and coordination between fire response and land management agencies has improved. However, there is still significant duplication of activities and conflicting strategic priorities across jurisdictions (Australian Senate, 2010).

Communication of forecasting information

The Bureau of Meteorology (BoM) has moved to a centralized model for aviation forecasting, but spot-fire forecasts for point locations are at the request of state government emergency services, who distribute this information to firefighters. This state-based approach allows specialized local knowledge of small-scale meteorological interactions that are not always captured by the modelling, adding significant value to forecasts.

The forecast provides an overview of the synoptic situation and describes any potential thunderstorm development and expected precipitation. Forecasts also include winds, wind gusts, temperature, dew point, fuel load and moisture. Dry thunderstorms are particularly relevant and are known to be the most common natural cause of bush fires in many parts of Australia.

The BoM is also responsible for producing the daily Forest & Grass Fire Danger Index data used to create a bushfire danger rating (Commonwealth of Australia, 2020).

2.3.2 Indigenous Engagement

To ensure effective bushfire management, response and policy making, it is critical to communicate with, and include Australia's Aboriginal and Torres Strait Islander community in all planning efforts. The 2019–2020 Australian bushfires brought renewed public interest in cultural burning practices through academic research publications and the news media (Williamson, *et al*, 2020). This has given prominence to the importance of Indigenous cultural burning practices as part of fuel management, however Indigenous involvement in policy making for fire preparation and management is largely absent from state and federal government approaches (Williamson, et. al, 2020). Case studies such as the World Heritage Listing of Kakadu National Park illustrate that a collaborative approach is possible and effective:

> In Kakadu, traditional ecological knowledge is being used in powerful combination with Western science to manage and monitor vital cultural and natural resources, leading to a dramatic enhancement of biodiversity and cultural values. (McGregor *et al*, 2010)

During the 2019–2020 Australian Bushfires, over 84,000 Indigenous people were estimated to live in bushfire-affected areas (Williamson *et al*, 2020). The complex land rights held by Indigenous people has led to alienation from traditional lands held under Native Title legislation and traditional Aboriginal law due to extensive bushfires (Williamson *et al*, 2020). Over one quarter of the Indigenous population of New South Wales and Victoria live in bushfire-affected areas, reinforcing the issue of under representation of Indigenous peoples in Australian bushfire response (Williamson *et al*, 2020).

Stakeholder engagement with Indigenous landowners in bushfire-affected areas is critical to managing and mitigating the effects of future bushfires. Following the example of Kakadu National Park, the space-based technologies discussed in this report should be employed collaboratively and in consultation with Indigenous stakeholders, incorporating their traditional ecological knowledge.

2.3.3   Fire Danger Ratings and Bushfire Warning Systems

Australia currently uses the Australian Fire Danger Rating System (AFDRS) and the newly launched Australian Warning System to communicate information about bushfires to the public (Appendix H).

Extensive reviews of the current AFDRS have concluded that the system is based on scientific models from the 1960s. The system is based only on weather parameters and does not account for environmental and human variables critical in predicting bushfire risk and behavior (Paton, 2006; Commonwealth of Australia, 2020). In 2014, the Australian Government committed to prioritize the development of a new AFDRS.

Following the Royal Commission (Commonwealth of Australia, 2020), a new, three-tiered national warning system was launched in December 2020. The warning levels are: Advice, Watch and Act and Emergency Warning. The fundamental difference in the new system is the consistency of symbols, colors and recommended actions to be used by all jurisdictions in response to natural hazards. The changes were designed to create greater public understanding and will be particularly useful for people at cross-border locations.

2.3.4   Communicating with Stakeholders

Effective and reliable communication is critical for successful bushfire management. Government, emergency service agencies and the public depend on communication services to make decisions during emergency situations. Communications infrastructure can be damaged or destroyed during bushfires with devastating consequences for evacuation and rescue efforts.

Early evacuation during bushfire events is widely agreed to be the most effective option to support emergency services' operations and prevent human injury and death (McLennan *et al*, 2018). While mandatory evacuation laws have been implemented in other countries,

Australian law and policy provides individuals with a choice to either evacuate or to stay and defend their property (Whittaker, Taylor and Bearman, 2020; McCaffrey, Rhodes and Stidham, 2014).

Bushfire warning messaging is considered to be one of the main factors contributing to action and evacuation behaviors (Strahan, Whittaker and Handmer, 2018; McLennan *et al*, 2018). To facilitate a timely and effective public response, warning messages must provide detailed information on fire behavior and recommended actions to those in danger (Paton, 2006).

Only three states have developed dedicated mobile applications to communicate bushfire emergency warnings and information for their geographic regions. New South Wales (Fires Near Me NSW), South Australia (AlertSA), and Victoria, (VicEmergency). Notably NSW Rural Fires Service (NSW RFS) has developed an application (Fires Near Me Australia) to geographically cover the entire country using data from participating fire agencies. All states and territories operate a fire agency website at a minimum, although most have also adopted social media as part of their communication strategy. See Appendix I for installation and review figures.

2.3.5 Emergency communication system directly between non-modified phones and satellites

Loss of communication was identified as a key issue in the 2019-2020 Australian bushfires (Commonwealth of Australia, 2020). Australia is not alone in this problem. The US Public Utilities Commission reported that 2017 Northern California wildfires left 85,000 wireless customers without service. This statistic included 11 to 15 Public Safety Answering Points, which are call centres responsible for answering emergency calls. A solution is needed to ensure communication channels remain resilient and accessible to the general public during an emergency.

This limitation could be overcome by eliminating the need for terrestrial mobile connections which are at a high risk of damage during bushfires. An innovative solution has been proposed, where mobile phones become satellite phones without additional hardware. This concept is being developed by a global enterprise Lynk (Lynk 2021), who have successfully broadcast emergency short message service (SMS) messages directly from satellite to a standard smartphone. The government could use this option to directly broadcast information to the community during emergency.

2.3.6 Education and Outreach

Currently, each Australian state is responsible for the management of the bushfire-related outreach and education (NSW Fire and Rescue, 2021; SA Country Fire Service, 2021; Fire Rescue Victoria, 2021). Different states provide information sessions, support for vulnerable communities and school-based programs to raise awareness about bushfire preparedness. A nationwide network of outreach programs and curriculum for bushfires preparedness has not been developed in Australia. Internationally, the USA has developed a federal fire organization operating under the Federal Emergency Management Agency (FEMA) (US Fire Administration, 2021).

2.3.7 Satellite imagery for effective communication of bushfire hazards through Virtual and Augmented Reality experiences

Bushfire hazard communication using complex scientific two-dimensional (2D) satellite imagery and other data may be challenging for non-trained viewers. However, the exponential increase of computer power in recent years enables satellite data to be reconstructed into3D visualisations (Rupnik *et al,* 2018). Such 3D imagery can be used to create Extended Reality (XR) experiences across various levels of immersion from Augmented Reality (AR) to Virtual Reality (VR). To access 3D information in real-time and on-demand with minimum latency, 5G connection through space assets is required. With the support of Artificial Intelligence (AI) and precise location via Global Navigation Satellite System (GNSS) data, tailored experiences can be created in the context of communicating bushfire hazard to affected communities, the general public, and first responders (Tuffley, 2019). VR/AR applications accessed through smart glasses and headsets could assist in emergency navigation, planning, training, empathy development and education. In addition, collaborative VR/AR environments can also play a significant role in efficient decision making during the emergency by providing incident commanders and first responders with a shared virtual interface (Turton, 2020).

2.3.8 The Disaster Charter

Several international rules govern technologies in remote sensing. Discussion is ongoing regarding how the International Charter on Cooperation to Achieve the Coordinated Use of Space Facilities in the Event of Natural or Technological Disasters (The Disaster Charter), may facilitate an additional early warning system and assist in predicting disasters.

The Disaster Charter covers not only remote sensing but also other space-based services of use in disaster situations (Mosteshar, 2016). The Charter mandates that there must be a clear description of space assets and extent of the coverage that they may provide so that it may be of effective assistance to countries in crisis.

The Charter has supported forty-nine wildfire cases since its initiation in 2000 by providing imagery from multiple satellites to assess the scope and analyze damage from wildfires (The International Charter Space and Major Disasters, 2020). As of August 2020, there had been 670 activations of the Charter, and approximately eight per cent of those activations were for wildfires (The International Charter Space and Major Disasters, 2020) (Appendix L).

3
Future State and Recommendations

Recommendation 1. Implement a national approach to bushfire management and cultivate a more connected stakeholder environment

We extend the recommendations of the Royal Commission (Commonwealth of Australia, 2020) and Bushfire Taskforce (Australian Space Agency, 2020) and discuss the need for implementation of a national bushfire policy and future research agenda, which addresses stakeholder identification and collaboration. We also discuss the need for a national program to foster innovation.

3.1 Establish a bushfire management organization at national level

In line with recommendations from the Royal Commission and Bushfire Taskforce (Commonwealth of Australia, 2020; Australian Space Agency, 2020), continued emphasis should be placed on establishing a National Department of Bushfire and Land Management. This new institution should 'assist with coordinating resources and developing standard operational systems across jurisdictions' as outlined in the *The incidence and severity of bushfires across Australia* inquiry (Australian Senate, 2010). Ideally, this organization should be an addition to the Australian Fire and Emergency Services Council (AFAC) model, assisting in the collection and processing of data obtained from space assets. The organization may also act as the financial vehicle for funding of space technology solutions, reducing the financial burden on state and territory departments.

This Department of Bushfire and Land Management would allow Australia to take a more strategic and coordinated approach to bushfires in future, and would act in an advisory capacity for state and territory organizations. This should be supported by a National Disaster Framework.

3.2   Generate interdisciplinary collaboration and communication within the Australian Government and invite early involvement of multi-jurisdictional agencies

A multi-jurisdictional, whole-of-government approach across all states and territories should be adopted, with co-ordination across a range of key stakeholders and multi-disciplinary agencies, including firefighting, engineering, sociology, psychology, geography and computer science, to ensure the smooth integration of space-based bushfire mitigation technologies.

3.3   Consult and collaborate with Indigenous Australian stakeholders to understand and incorporate cultural land practices

Following the example of Kakadu National Park (Section 2.3.2), the technologies and strategies discussed in this report should be employed collaboratively and in consultation with Indigenous stakeholders, incorporating their traditional ecological knowledge. Effective engagement and consultation in government policymaking demonstrates recognition of traditional knowledge and allows greater control by Indigenous peoples as custodians of their traditional lands. This collaborative approach 'holds great promise for better management of the world's natural resources' and represents the best way forward in effective Australian land and bushfire management (McGregor *et al*, 2010: 721).

3.4   Promote community level and behavioral resilience

Australian governments (local, state and federal) should adopt policies to empower individual communities to build resilience. Disaster resilience varies greatly at a local level, and those best placed to prepare for fire are communities themselves. An Australian policy of community-level resilience could be modelled on California's new Wildfire and Forest resilience Action Plan, which concentrates on assisting communities to identify their unique set of needs, values, risks, and capacities and to develop a common framework to facilitate comprehensive local plans to enhance resilience (California Department of Water Resource, Public Affairs Office, 2021). Community-level resilience planning could be further informed by publicly available bushfire risk maps derived from satellite data, with granular detail at the household level (See Recommendation 5).

3.5   Cooperate internationally and promote adherence to international agreements related to fire management, research and development

The national organization should be responsible for developing international cooperative relationships and resource sharing partnerships.

It is recognised that bushfires are a global issue. International partnerships and arrangements that currently exist between Australia and its counterparts provide an opportunity to focus on the research, development and operationalization of space technologies for better bushfire management. Developing partnerships across the international community will enable Australia to understand the requirements and timelines to implement similar missions. It will also be important to foster relationships and collaborations with the broader space community to enable future steps.

The International Charter for Space and Major Disasters should be adopted in Australia to enable effective worldwide collaboration and shared responsibility during disasters.

3.6   Strengthen International partnerships to accelerate research and development

The Canadian Space Agency (CSA) is planning its WildFireSat mission to enable precise monitoring of wildfires, smoke and air quality on a daily basis (Canadian Space Agency, 2020). Canada's WildFireSat mission is expected to combat similar bushfire challenges as are faced in Australia. The mission is expected to launch in 2026. WildFireSat has already attracted Letters of Interest from NASA,

NOAA, USFS (US Forest Service), South Africa and the UN's FAO (Food and Agriculture Organization), highlighting a desire for usage beyond Canada.

Since signing the Memorandum of Understanding with the CSA in 2018, the Australian Space Agency has been engaging closely to identify opportunities for Australia, which could facilitate wider collaboration. The Australian government and Australian Space Agency should explore opportunities for involvement in the design, development, implementation, and co-ownership of WildFireSat.

Keeping informed of the progress of the mission, and exploring the opportunity to assist, will also help Australia understand the requirements, challenges and timelines that would need to be accommodated if the mission was to be replicated by Australia in the future.

3.7   Implement bushfire education programs across various levels of government

Implementation of official national programs across primary, secondary and tertiary education would ensure a consistent understanding of natural hazards. Effective bushfire education may promote community adhesion in a safe and equal manner in response to an emergency. Although responsibility for education is separated by jurisdiction, the implementation and its messaging should be nationally consistent.

Educational programs could be supported through collaboration with bushfire experts, engineers and education consultants to ensure the Australian school curriculum teaches students the scientific principles of bushfire prevention and mitigation. This includes education on the Fire Danger Ranging System (FDRS), Australian Warning System (AWS), and how to develop and execute a bushfire action plan (Appendix G, H).

Internships and work experience opportunities for students in space agencies and governments could be a secondary approach to inspire the younger generation (South Australian Space Industry Centre (b), 2021).

3.8   Foster collaboration for space innovation

The national organization should foster a culture of research and innovation to empower industries and start-ups to respond to some of the more complex social and environmental challenges of our time, including bushfires (Appendix M). To facilitate this:

- support should be proactively offered by government and public-private co-operation on leading challenges prioritized
- research centers for innovative technologies & solutions (robots, unmanned ground and air vehicles, satellite-based solutions, earth observation and remote sensing, AI-based image, and video processing) should be established
- the organization should capitalize on opportunities to learn from or be a part of international bushfire technology initiatives
- space innovation should be promoted through primary, secondary and tertiary education curriculums and programs, including hackathons or innovation sessions.

Recommendation 2. Improve data infrastructure to support a centralized data platform. This should incorporate multiple data sources to better support a longer-term risk and prediction model

Cloud infrastructure technologies offer a solution for storing large quantities of data, such as those from satellites and other space-based sensors, while also allowing the processing power required to derive meaning from such data. Such a solution could allow the training and retraining of data for a prediction model to be completed within hours, if not minutes, rather than days and weeks, as with standalone infrastructure. It may also offer opportunities for a significantly longer prediction window, which would be valuable in order to allow timely information for bushfire response teams, and rapid response deployment to impacted areas.

Current efforts should be focused on consolidating existing data around fuel moisture content and fuel loads from current space assets. The centralization and processing of the data from multiple sources including satellites, Geographic Information Systems (GIS), ground-based sensors, and aerial mapping, allows relevant agencies and communities to share in a more comprehensive single-view risk and prediction model.

We assume that if a centralized bushfire or disaster agency is established the organization would be the owner of the data platform. Later stages could involve providing access and insights to other community stakeholders. The data platform would store all the data and this in itself would require external access, however the insights generated from the platform should sit in an open source or application programming interface (API) environment where those impacted, such as firefighters and communities, can access as required. Providing open data also supports private enterprise and the community to use the data feeds to create innovative new and useful products.

Recommendation 3. Deploy lightning specific sensors on satellites

Given that eighty-two per cent of bushfires are started by lightning strike, the ability to predict when a strike will occur is a critical component of any fire prediction system.

Existing satellite data relating to fuel load, fuel moisture content, and weather observations can be coupled with the lightning data. The addition of sensors that focus on lightning formation, such as infrared cloud top temperature, cloud-to-cloud and intra-cloud lightning would be an effective way of predicting fire prior to ignition for the shorter term.

Investigating the feasibility of a dedicated sensor covering the Australian region to indicate the presence of lightening, such as a near infrared optical transient detector, would assist, noting that the Geostationary Lightning Mapper on GOES-16 covers the Americas.

Low-cost alternatives to supplement data at specific locations through complementary ground-based solutions should also be considered. Commercial companies like Biral, Earthnetworks and Vaisala offer ground-based sensor solutions. These low-cost alternatives could be used in specific locations to supplement data requirements, with commercial apps for end users, monitoring of networks and delegation of management of assets. These ground-based sensors can interact with space assets to provide an improved overview, allowing resources to be focused on specific priority locations.

Recommendation 4. Use space assets to detect vegetation, topography and moisture content to identify 'at-risk' areas where lightning could spark bushfires

LiDAR and short-wave infrared sensors have been identified as an option to locate and map large areas of fuel content for bushfires. Using LiDAR 3D mapping an accurate picture of the terrain can be constructed and uphill dry vegetation can be highlighted. The data can be filtered and integrated to develop a GIS that can create a map of areas for pre-emptive hazard reduction burning. This will combine the information gathered from satellites along with the locations of human settlements to highlight areas that will affect and impede human life in the event of fire (Figure 3).

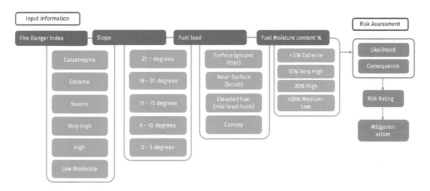

Figure 3: Fire Mitigation Risk Assessment Process

Recommendation 5. Enhance end-to-end bushfire response mitigation to include predictive analysis and communications

5.1 Develop a real-time nationwide risk assessment platform and state-wide dynamic risk profiles

We recommend development of a national risk map as a 'single point of truth' for fire risk across the country, with detail at the household level. This single national platform must be reliable and accessible by multiple channels (including mobile application, SMS, e-mail, etc). The platform will provide accurate, real-time fire information to individuals and organisations, allowing them to make more informed decisions both in preparation for and in response to bushfire.

Bushfire risk management should be implemented and driven at federal level to provide information to all stakeholders. It should align with the National Disaster Risk Reduction Framework that was created by the Australian Government Department of Home Affairs (Australian Government 2018).

5.2   Use the risk map and tools for state and local bushfire impact mitigation plans

Location-specific risk-based assessments informed by predictive fire behavior and risk modeling tools should replace generic area-based targets (Russell-Smith, McCaw and Leavesley, 2020). More focused information can be used to better target fuel reduction burns, zoning, awareness raising, and budgetary resources. The real-time risk map is also able to be used by emergency services to communicate alert levels and issue evacuation warnings, enabling cohesive state and local government messaging in critical emergency situations.

5.3   Use the risk map to develop and foster a culture of resident-level fire plans

A community empowerment model (Recommendation 1.4) requires individuals to have access to real-time, user-friendly risk maps and profiles to inform their Bushfire Survival Plans, evacuation plans, and to protect themselves and their communities.

Recommendation 6. Use space-based life support system technologies and materials for enhancing firefighting protective equipment

6.1   Repurpose ideas and principles of the Astronaut Health Monitoring System (AHMS) to develop advanced firefighting suits

Firefighting suits modelled after those worn by astronauts, incorporating sensor-based technologies, such as Industrial Internet of Things and an on-board computer, would enable real-time incoming and outgoing data feeds, increasing environmental situational awareness for both the firefighter and fire management services (Appendix F) (Grant 2016). Sensor-based technologies to monitor firefighter health metrics would enable effective firefighting workforce management and support a sustainable bushfire response (Commonwealth of Australia, 2020:184). An on-board computer housed in a fire-retardant alloy would ensure operational capability in localized settings, minimizing communication constraints. This could be further refined with addition of a distributed data-centric model enabling AI technologies (Grant 2016)(Figure 4).

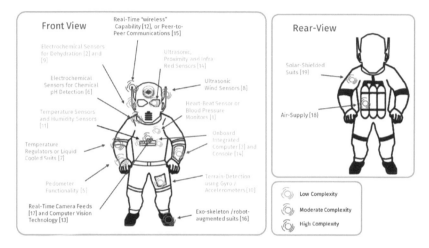

Figure 4: Detailed components of the firefighter suit concept design

6.2   Use thermal protective materials from spacecraft for hardening of fire trucks for extreme temperatures

Reimagining firefighting vehicles to incorporate space shuttle thermal technology and life support systems may be key to mitigating injury and fatalities during a burn-over or entrapment scenario (Penney, *et al*, 2020) (Figure 5). The Australian Standard 1530.8.2 [Methods for fire tests on building materials, components and structures; Part 8.2] was identified as the benchmark (Standards Australia, 2018, Penney, et al., 2020). To investigate technology transfer from the space industry to building and vehicle materials, it is proposed that NASA's Thermal Protection Materials Branch patents be released to explore appropriate materials to enhance firefighting vehicle protection systems (NASA, 2019). Space shuttle exterior design to deflect heat should also be taken into consideration (Appendix F) (Sivolella, 2017).

The overall cost and fragility of the thermal tiles may make them impractical for most Earth-bound applications. A more recent advanced thermal coating technology, based in part on animal bone charcoal may be more practical (ENBIO, 2020). This coating is used on a heatshield and other exposed surfaces of the Solar Orbiter mission. Launched in 2020, it will approach the Sun at around a quarter of the distance from the star to Earth. Solar Orbiter will experience temperatures averaging 500°C due to its proximity, and this coating technology helps to protect exposed surfaces.

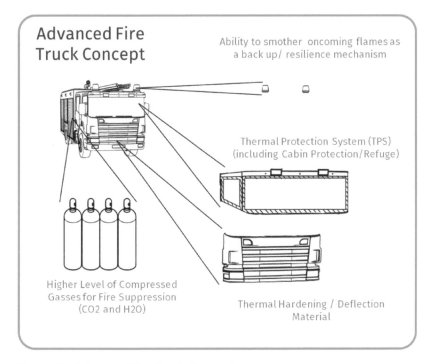

Figure 5: Advanced Fire Truck Concept

6.3   Develop advanced thermal sheltering mechanisms for fire refuge locations

Mitigation of fatality and injury during a bushfire cannot solely rely on early evacuation on extreme fire danger index days (BOM, 2021). It is proposed to have thermal sheltering facilities and/or underground bunkers with Environmental Control and Life Support Systems (ECLSS) that will sustain life for a designated period of time during a bushfire (Schauble, 2013; Thangavelu, *et al*, 2020). The key attributes would be to maintain a breathable atmosphere through a regenerative system that included water, temperature, humidity control and fire suppression and detection (Sivolella, 2016). Appropriate power sources with redundancies due to the high risk of power failure during a bushfire should also be considered.

Recommendation 7. Fortify communications infrastructure to increase bushfire resiliency

7.1 Use proactive fire suppression techniques to fortify communications infrastructure

Reliable communications infrastructure is critical to stakeholders during bushfires and in recovery. Critical communications infrastructure connecting to satellite and land information systems should be built or retrofitted with capabilities, some of which were originally developed by space programs, to facilitate continued operation despite extreme heat, including:

- Heat insulated housings for non-structural components and hardware
- A more evenly distributed power network to mitigate power loss (Smart Grid)
- Fire retardant chemical reservoirs connected to sensor activated dispersion systems
- Autonomous vegetation maintenance systems to maintain asset fire break zones

The ground infrastructure network linking with space communication networks must have built-in redundancy to compensate for fire damage to primary and secondary towers. Using thermal coating technology like that used on spacecraft would also ensure the thermal hardening of critical infrastructure (Enbio, 2020).

7.2 Implement a direct satellite-to-mobile phone communication system for emergency communications

An emergency space communication system would allow a direct link between non-modified smartphones and satellites, which work as space cell towers. Although not yet in use in Australia, this concept has been shown to work by Lynk (2021). The satellite constellation in Low Earth Orbit would grant full 2G communication coverage across Australia.

The change in connection between the mobile-device to towers or base stations, to a satellite uplink, would occur automatically when the phone has insufficient coverage.

The phone will connect directly to the satellite signal, and will provide SMS connectivity directly to the smartphone. The government can also use this to broadcast information to an application, where the data can be displayed in a user-friendly way. The application can also have an SMS service so the users can easily interact with others. There is ongoing research to provide a 4G link (See Appendix K).

Recommendation 8. Implement a national Australian Fire Danger Rating System and improved Bushfire Warning System, and a national bushfire emergency application to support these systems

8.1   Accelerate the development and implementation of a National Australian Fire Danger Rating System

The proposed Australian Fire Danger Rating System is not scheduled to be implemented until 2023 leaving Australia vulnerable to catastrophic bushfire seasons in interim years. The federal government should provide additional resources to support accelerated completion of field testing, design, and consultation phases.

8.2   Enhance the Australian Warning System

The new Australian Warning System offers significant improvements on the previous system. However, confusion remains around the second level warning: 'Watch and Act'. The wording leaves people questioning if the advice is to simply monitor the situation or requires them to act, which are fundamentally different behavioral responses. A clearer warning descriptor should be used to convey a precise message (Anderson-Berry *et al*, 2018).

8.3   Use the consolidated data platform to implement a national bushfire emergency application

We propose analyzing and sharing critical data from the centralized data platform (Recommendation 2) to enable real time communication to all stakeholders across jurisdictions. A national emergency and bushfire alert application linked to the centralized data platform would ensure consistent, timely messaging and education to the public, as well as rapid firefighting response. This will also eliminate the duplication of effort by state and territory fire agencies to develop and maintain independent applications that are functionally similar.

4

Conclusion

As a group of individuals brought together by our common passion for space, it will come as no surprise that our team project report identified, and recommended, several ways that space assets and technologies can be leveraged to enhance bushfire management in Australia.

While some of our recommendations are similar to those from previous work in this area, we hope we have also provided some novel and innovative ideas for stakeholders to consider, as they decide how best to deploy their resources in this battle to save lives, property, and the environment.

As a team, we have learned a lot from each other during this process. Our Australian hosts have openly shared their first-hand experiences of the effects of bushfires in their country. Those of us from other distant lands responded by immersing ourselves in the literature, absorbing new knowledge and offering fresh ideas and alternative perspectives. Together our SHSSP21 team proudly, but humbly, encourages our readers to consider carefully and hopefully pursue the recommendations presented in this report. By following the Interdisciplinary, International and Intercultural approach described above, we hope that some of the devastation seen during past bushfire events could be avoided in future fire seasons.

5

References

Anderson-Berry, L, Achilles, T, Panchuk, S, Mackie, B, Canterford, S, Leck, A and Bird, D, 2018. Sending a message: How significant events have influenced the warnings landscape in Australia. *International Journal of Disaster Risk Reduction*, [online] 30, 5–17. Available at: <https://www.sciencedirect.com/science/article/pii/S2212420918302760> [Accessed 11 February 2020].

Australian Government. 2018. National Disaster Risk Reduction Framework. *Department of Home Affairs*. https://www.homeaffairs.gov.au/emergency/files/national-disaster-risk-reduction-framework.pdf

Australian National University 2020. Eyes in space to spot bushfire danger zones. 4 March 2020. https://www.anu.edu.au/news/all-news/eyes-in-space-to-spot-bushfire-danger-zones

Australian Senate, 2010. *The Incidence and Severity of Bushfires Across Australia.* Canberra: Select Committee on Agricultural and Related Industries. 13 August 2010.

Australian Space Agency 2020. Bushfire Earth Observation Taskforce: Report on the role of space-based Earth observations to support planning, response and recovery for bushfires. https://www.industry.gov.au/sites/default/files/2020-12/bushfire-earth-observation-taskforce-report.pdf.

Bauer, P, Thorpe, A & Brunet, G, 2015. The quiet revolution of numerical weather prediction. *Nature,* 525, 47–55. https://doi-org.ezproxy.csu.edu.au/10.1038/nature14956.

Bureau of Meteorology (BoM). 2021. Fire Weather Knowledge Centre. http://www.bom.gov.au/weather-services/fire-weather-centre/index.shtml [accessed 15 February 2021]

Borchers Arriagada N, Palmer AJ, Bowman DMJS, Morgan GG, Jalaludin BB, Johnston FH, 2020. Unprecedented smoke-related health burden associated with the 2019–20 bushfires in eastern Australia. *Med. J. Aust.* Med J Aust 2020; 213 (6): 282–283. https://doi.org/10.5694/mja2.50545

Bureau of Meteorology 2020. State of the Climate 2020. http://www.bom.
gov.au/state-of-the-climate/documents/State-of-the-Climate-2020.pdf.

Bureau of Meteorology. *What is El Niño and what might it mean for Australia?*
[Online]. Available: http://www.bom.gov.au/climate/updates/articles/
a008-el-nino-and-australia.shtml [Accessed 11 February 2021].

Bush, E, 2015. Breathing wildfire smoke can have serious health
consequences. *Michigan State University Extension.* https://www.canr.
msu.edu/news/breathing_wildfire_smoke_can_have_serious_health_
consequences [Accessed 11 February 2021].

California Department of Water Resource, Public Affairs Office. 2021.
California's Wildfire and Forest Resilience Action Plan. https://fmtf.fire.
ca.gov/media/cjwfpckz/californiawildfireandforestresilienceactionplan.pdf

Canadian Space Agency, 2020. WildFireSat: Enhancing Canada's ability to
manage wildfires. https://asc-csa.gc.ca/eng/satellites/wildfiresat/default.
asp [Accessed 16 February, 2021]

California Natural Resources Agency (CNRA). 2021. Governor's Task Force
Outlines Actions to Reduce Wildfire Risk, Improve Health of Forests and
Wildlands. https://resources.ca.gov/Newsroom/Page-Content/News-List/
Governors-Task-Force-Outlines-Actions-to-Reduce-Wildfire-Risk

Commonwealth of Australia 2020. Royal Commission into National Natural
Disaster Arrangements Report. https://naturaldisaster.royalcommission.
gov.au/system/files/2020-11/Royal%20Commission%20into%20
National%20Natural%20Disaster%20Arrangements%20-%20Report%20
%20%5Baccessible%5D.pdf.

Cook, D, 2020. Open data shows lightning, not arson, was the likely cause
of most Victorian bushfires last summer, *The Conversation.* https://
theconversation.com/open-data-shows-lightning-not-arson-was-the-
likely-cause-of-most-victorian-bushfires-last-summer-151912

CSIRO. 2020. Climate and Disaster: Technical Reports [online] https://www.
csiro.au/en/Research/Environment/Extreme-Events/Bushfire/frontline-
support/report-climate-disaste-resilience [Accessed 22 January 2021].

Enbio. 2020. Solar Orbiter. http://www.enbio.eu/solar-orbiter/ [accessed 18
Feb 2021]

Eriksen, C and Gill, N, 2010. Bushfire and everyday life: Examining the
awareness-action 'gap' in changing rural landscapes. Geoforum, Volume
41, Issue 5, Pages 814–825. https://www.sciencedirect.com/science/
article/pii/S0016718510000576

Foverskov, L, 2020. Preventing Forest Fires with Smart Technology. https://
orangematter.solarwinds.com/2020/04/13/preventing-forest-fires-with-
smart-technology/#:~:text=Over%20th%20last%20few%20years,and%20
effectively%20respond%20to%20bushfires [Accessed 15 February, 2021]

Grant, C, 2016. *The Future American Fire Fighter.* Fire Protection Research Foundation and National Fire Protection Association. https://www.nfpa.org/-/media/Files/News-and-Research/Resources/Research-Foundation/Current-projects/Smart-FF/SmartFFRealizingTheVision.ashx

Jain, P, Coogan, SCP, Subramanian, SG, Crowley, M, Steve Taylor & Flannigan, MD, 2020. A review of machine learning applications in wildfire science and management. *Environmental Reviews.* https://doi.org/10.1139/er-2020-2019.

Lynk. 2021. *Lynk—Our Technology* Available at: lynk.world/our-technology [Accessed 16 February 2021].

McCaffrey, S, Rhodes, A and Stidham, M, 2015. Wildfire evacuation and its alternatives: perspectives from four United States' communities. *International Journal of Wildland Fire,* [online] 24(2), p.170. Available at: <https://www-publish-csiro-au.access.library.unisa.edu.au/wf/WF13050> [Accessed 10 February 2020].

McGregor, S, Lawson, V, Christopherson, P, *et al,* 2010. Natural and Cultural Resources in Australia's World Heritage-listed Kakadu National Park. Human Ecology 38:721–729.

McLennan, B and Eburn, M, 2012. *Exposing hidden-value trade-offs: sharing wildfire management responsibility between government and citizens.* [online] International Journal of Wildland Fire. https://www.publish.csiro.au/wf/Fulltext/WF12201 [Accessed 15 February 2021].

Mäkelä, 2006. Comparison between lightning data and cloud top temperatures in Finland, Semantic Scholar, Corpus ID: 208876841.

Marshall, S, 2021. Marshall Government to send satellite to space. https://www.premier.sa.gov.au/news/media-releases/news/marshall-government-to-send-satellite-to-space.

Mosteshar, S, 2016. *Regulation of remote sensing by satellites.* Routledge Handbook of Space Law, p.144

NASA. 2019. *Thermal Protection Materials* Branch: Patents. https://www.nasa.gov/centers/ames/thermal-protection-materials/patents.html [accessed 13 February 2021]

National Oceanic and Atmospheric Administration 2017. Flashy First Images Arrive from NOAA's GOES-16 Lightning Mapper. [online] Available at: <https://www.nesdis.noaa.gov/content/flashy-first-images-arrive-noaa%E2%80%99s-goes-16-lightning-mapper> [Accessed 12 February 2021].

National Oceanic and Atmospheric Administration 2018. Tracking Lightning from Space: How Satellites Keep You Safe During Thunderstorms. [online] Available at: <https://www.nesdis.noaa.gov/content/flashy-first-images-arrive-noaa%E2%80%99s-goes-16-lightning-mapper> [Accessed 12 February 2021].

New South Wales (NSW) Fire and Rescue. 2021. Uniforms and equipment. https://www.fire.nsw.gov.au/page.php?id=164

Orange, 2017. 5 digital technologies to help fight wildfires. Available at: https://www.orange-business.com/en/magazine/5-digital-technologies-to-help-fight-wildfires [Accessed 15 February 2021]

Paton, D, 2006. *Warning Systems: Issues and considerations for warning the public.* [online] Bushfire Cooperative Research Centre. Available at: <https://www.bushfirecrc.com/sites/default/files/managed/resource/paton-bushfire-warnings_wcover_1.pdf> [Accessed 10 February 2021].

Penney, G, Habibi, D, Cattani, M, 2020. *Improving firefighter tenability during entrapment and burnover: An analysis of vehicle protection systems.* Fire Safety Journal, vol 118, 103209. https://www.sciencedirect.com/science/article/abs/pii/S0379711220301922?via%3Dihub

Peters, A, 2020. *This tool is mapping every tree in California to help stop megafires.* Fast Company. https://www.fastcompany.com/90549160/this-tool-is-mapping-every-tree-in-california-to-help-stop-megafires [accessed 10 Feb. 2021]

Rupnik, E,, Pierrot-Deseilligny, M. and Delorme, A., 2018. 3D reconstruction from multi-view VHR-satellite images in MicMac. ISPRS Journal of Photogrammetry and Remote Sensing, 139, pp.201–211.

Russell-Smith, J; McCaw, L. and Leavesley, A. 2020. Adaptive prescribed burning in Australia for the early 21st Century—context, status, challenges. *International Journal of Wildland Fire,* 29, pp.305–313. https://www.publish.csiro.au/wf/wf20027

Schauble, J, 2013. *How Should We Shelter from Intense Bushfires?* [e-journal] *Wildfire,* vol. 22, no. 1, pp. 14–20. https://search.ebscohost.com/login.aspx?direct=true&AuthType=cookie,ip,shib&db=bth&AN=85193811&scope=site [accessed 11 February 2021]

Sivolella, D, 2016. Life Support Systems of the International Space Station [chapter] in *Handbook of Life Support Systems for Spacecraft and Extraterrestrial Habitats* [e-book] Switzerland, Springer, Cham https://doi.org/10.1007/978-3-319-09575-2 [Accessed 15 February 2021]

Sivolella, D, 2017. *The Space Shuttle Program* [e-book] Switzerland, Springer, Cham. https://doi.org/10.1007/978-3-319-09575-2 [Accessed 14 February 2021]

Standards Australia. 2018. AS 1530.8.2 Methods for fire tests on building materials, components and structures—Part 8.2 Tests on elements of construction for buildings exposed to simulated bushfire attack—large flaming sources, Standards Australia, Sydney, N.S.W. https://www.standards.org.au/standards-catalogue/sa-snz/building/fp-018/as-1530-dot-8-dot-2-colon-2018

Strahan, K, Whittaker, J and Handmer, J, 2018. Predicting self-evacuation in Australian bushfire. *Environmental Hazards*, [online] 18(2), pp.146–172. Available at: <https://www-tandfonline-com.access.library.unisa.edu.au/doi/pdf/10.1080/17477891.2018.1512468?needAccess=true> [Accessed 10 February 2020].

Thangavelu, M, Abdrakhman, U and Zhang, X, 2020. *Outer Space Activities and City Evolution.* [e-journal] AIAA, vol. 4059, pp. 1–21 http://spacearchitect.org/pubs/AIAA-2020-4059.pdf [Accessed 16 February 2021]

The International Charter Space and Major Disasters. 2020. *Fire monitoring through the Charter.* https://disasterscharter.org/web/guest/-/fire-monitoring-through-the-charter

Tian, S, Van Dijk, A, Tregoning, P & Renzullo, L, 2019. Forecasting dryland vegetation condition months in advance through satellite data assimilation. *Nature Communications*, 10. Tuffley, D, 2019. *Virtual tools, real fires: how holograms and other tech could help outsmart bushfires.* [online] The Conversation. Available at: <https://theconversation.com/virtual-tools-real-fires-how-holograms-and-other-tech-could-help-outsmart-bushfires-126830> [Accessed 16 February 2021]. https://doi.org/10.1038/s41467-019-08403-x.

Turton, B, 2020. AROC by BadVR—Revolutionizing First Responder Technology. [online] Medium. Available at: <https://medium.com/badvr/aroc-by-badvr-revolutionizing-first-responder-technology-f8d302646c57> [Accessed 11 February 2021].

University of Southern Queensland 2021. Tech used in space to help detect fires in Australia. https://www.usq.edu.au/news/2020/02/space-tech-detect-fires-aus.

United States Office of Outer Space Affairs 2012. United Nations Programme on Space Applications. United Nations publication ST/SPACE/52/Rev 1 V.12-55442—September 2012. https://www.unoosa.org/pdf/publications/ST_SPACE_52_Rev1.pdf

Whittaker, J, Taylor, M and Bearman, C, 2020. Why don't bushfire warnings work as intended? Responses to official warnings during bushfires in New South Wales, Australia. *International Journal of Disaster Risk Reduction*, [online] 45. Available at: <https://www-sciencedirect-com.access.library.unisa.edu.au/science/article/pii/S2212420919304625> [Accessed 10 February 2020].

Williamson, B, Markham, F, Weir, JK, 2020. *Aboriginal Peoples and the Response to the 2019–2020 Bushfires. Centre for Aboriginal Economic Policy Research.* CAEPR Working Paper 134. Canberra: ANU College of Arts & Social Sciences.

Yebra, M, Quan, X, Riaño, D, Larraondo, PR, Dijk, AIJMV & J Cary, G, 2018. A fuel moisture content and flammability monitoring methodology for continental Australia based on optical remote sensing. *Remote Sensing of Environment*, 212, 260–272. https://doi.org/10.1016/j.rse.2018.04.053

6
Appendices

Appendix A.   Physical Fire Prediction Model

Figure 1 shows the key factors involved in the physical system of relevance to fire prediction, including the scope-boundary between prediction and mitigation. The system focusses on those aspects that can be detected by space assets. Space assets provide data that can be input into bushfire prediction models.

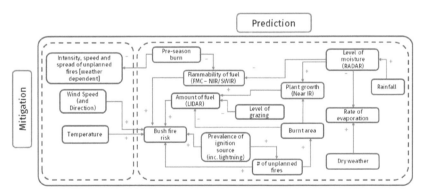

Appendix Figure A: Physical Fire Prediction Model

Appendix B.   Satellites used in weather, moisture, lightning, and fuel load detection

Existing satellites of use for the prediction of fires in natural landscapes are captured in Appendix Tables 1 & 2:

Appendix Table B1: Current satellites of relevance to bushfire prediction

Satellite	Orbit	Usefulness for bushfire prediction	Limitations
Himawari	• In GEO orbit • 8 launched so far • Himawari-8 is the only one operational now • Himawari-9 is to be launched after the 8th is retired	• Japanese weather satellites • Includes coverage of the Australian region • Near real-time data is available for the public • Directly feeds the project 'Mapping Major Bushfires in Australia using Near Real-time Himawari-8 Satellite Imagery'	• Low resolution • Data is useful only after bushfires take place
EarthCARE	• Sun-Synchronous • Near-Polar Orbit • LEO of 400km • Orbit cycle of 25 days	• Visible light LiDAR, with depolarising channel for moisture. This can create a 3D map of the topography, canopy tops for fuel loads and potentially the moisture content of the vegetation. • A Multi-Spectral Imager (MSI) will have channels in the VNIR, SWIR and thermal infrared. Can see through cloud cover, to show heat signatures and moist vegetation (EarthCARE—Earth Online, 2021).	• Launch end of 2021 • LiDAR swaths across Australia are limited and do not cover the full bushland. • Appendices Image X • LiDAR in visible light spectrum cannot see through cloud cover.

CALIPSO	LEO 685km Period 16 days	Main instrument: CALIOP, a backscattering LIDAR comprising two channels (532 nm and 1064nm) equipped with a telescope 1m diameter.	Mission extended to 2023
MODIS	LEO 705 km, 10:30 a.m. descending node (Terra) or 1:30 p.m. ascending node (Aqua), sun-synchronous, near-polar, circular;	• Vegetation indices; • Thermal anomalies, fire; • Surface reflectance; -Land surface temperature/emissivity; -Land cover dynamics (global vegetation phenology);	Design life: 6 years; MODIS cannot observe the surface when cloud cover is present. The visible bands of MODIS are only used during the day, when reflectance is the dominant mechanism for detecting ice and snow. The thermal bands are used during both day and night to measure sea ice surface temperature by emittance.
ICESat-2	• Near Polar LEO frozen orbit. • Orbit cycle of 91 days (ICESat-2—eoPortal Directory—Satellite Missions, 2021)	• ATLAS Laser altimeter with a mission goal to detect canopy depth as an estimate of Biomass. With 3m accuracy at 1km spatial resolution (Science | ICESat-2, 2021). • Polarized for Ice sheet moisture/mass detection. Potential to be repurposed for vegetation moisture. It uses visible light with a wavelength of 532nm which is green shifted (NASA Goddard, 2018).	• Only planned for a three-year lifetime. 2 years and 4 months elapsed. • Long orbit cycle.

Appendix Table B2: Satellites used in Earth observation and real-time wildfire detection

Name	Orbit	Usefulness for bushfire prediction
GOES-16	• GEO with coverage over the western hemisphere	• Contains the instrument GLM • Effective in detecting lightning • Lightning often is the ignition source for wildfires/bushfires • Therefore, it helps predicting wildfires/bushfires • However, it doesn't have coverage over the Australian region
Sentinel 1	• 2 polar-orbiting satellites • Period 12 day repeat cycle, • LEO 693 km	• Not activated all time, Cover mainly North Hemisphere • Limited number of measurements 111 since 2015 • Synthetic Aperture Radar (SAR)
Sentinel 2	• 2 polar-orbiting satellites • Sun synchronous, LEO 786km • Period 10 day repeat cycle whit the first spacecraft, reduced to 5days with both satellites	• MultiSpectral Instrument • High resolution, multispectral images
Sentinel 3	• 2 in-orbit • Sun synchronous orbit, 814km Repeat Cycle 27 days Swath Width OLCI 1270km SLSTR 1420 km	Global coverage vegetation products • OLCI (Ocean and Land Color Instrument) • SLSTR (Sea and Land Surface Temperature Radiometer) • SRAL (Synthetic Aperture Radar Altimeter) • MWR (Microwave Radiometer) • DORIS • LRR (Laser Retroreflector) • GNSS (Global Navigation Satellite System) Mission objectives: • Land-use change monitoring • Forest cover mapping • Fire detection • Weather forecasting • Measuring Earth's thermal radiation for atmospheric applications

Sentinel 5P–Oct 2017	• 1 near polar sun-synchronous orbit • Sun synchronous, MEO 2600km Period 15 days	• The state-of-the-art Tropomi instrument to map a multitude of trace gases such as nitrogen dioxide, ozone, formaldehyde, sulphur dioxide, methane, carbon monoxide and aerosols– all of which affect the air we breathe and therefore our health, and our climate. • A single instrument which is a UV-VIS-NIR-SWIR spectrometer
Landsat 7	Polar, sun-synchronous 16 day repeat cycle	• Earth-observing satellite. • PAN band 15m spatial resolution • Thermal IR with 60m Spatial resolution • October 2008, USGS made all Landsat 7 data free to the public (all Landsat data were made free in January 2009 • 16-day repeat cycle
Landsat 8	sun-synchronous, near-polar orbit 16 day repeat cycle	• Operational Land Imager (OLI) and the Thermal Infrared Sensor (TIRS) instruments • 30m spatial resolution for OLI • PAN 15m • Thermal Infrared Sensor 100m https://www.usgs.gov/media/images/landsat-photo-wallow-fire-arizona Image Dimensions: 3000 x 3000 Date Taken: TUESDAY, JUNE 7, 2011 Location Taken: Wallow, AZ, US

Landsat 9	To orbit Earth every 16 days in an 8-day offset with Landsat 8	• Operational Land Imager (OLI) and the Thermal Infrared Sensor (TIRS) instruments • Launch date September 2021 replace Landsat 7 • Resolutions: • OLI 30. • Pan 15m • IR 100m • Earth every 16 days in an 8-day offset with Landsat 8 • https://www.usgs.gov/core-science-systems/nli/landsat/landsat-9?qt-science_support_page_related_con=0#qt-science_support_page_related_con

Appendix B References:

Calipso.cnes.fr, 2021 *Calipso | Le site du Centre National d'Etudes Spatiales* [online] Available at: <https://calipso.cnes.fr/fr/CALIPSO/Fr/index.htm> [Accessed 15 February 2021]

Earth.esa.int. 2021. *EarthCARE—Earth Online.* [online] Available at: <https://earth.esa.int/eogateway/missions/earthcare> [Accessed 11 February 2021].

Earth.esa.int. 2021. *ICESat-2-eoPortal Directory-Satellite Missions.* [online] Available at: <https://earth.esa.int/web/eoportal/satellite-missions/i/icesat-2> [Accessed 15 February 2021].

Earth.esa.int, 2021 *Sentinel-3—ESA EO Missions—Earth Online—ESA* [online] Available at: <https://earth.esa.int/web/guest/missions/esa-eo-missions/sentinel-3> [Accessed 15 February 2021].

ESOV 2021. ESOV—the Earth Observation Swath and Orbit Visualisation tool. ESA.

Goes-r.gov. n.d. *GLM. GOES-R Series.* [online] Available at: <https://www.goes-r.gov/spacesegment/glm.html> [Accessed 12 February 2021].

Icesat-2.gsfc.nasa.gov. 2021. *Science | ICESat-2.* [online] Available at: <https://icesat-2.gsfc.nasa.gov/science> [Accessed 8 February 2021].

Leblon, B, Bourgeau-Chavez, L and San-Miguel-Ayanz, J, 2012. Use of Remote Sensing in Wildfire Management. Sustainable Development-Authoritative and Leading Edge Content for Environmental Management,.

NASA Goddard, 2018. *ICESat-2 By the Numbers: 532.* [video] Available at: <https://www.youtube.com/watch?v=O4yeiZ0s_7g&feature=emb_rel_end> [Accessed 15 February 2021].

Sentinel.esa.int, 2021 *Sentinel Online –ESA—Sentinel* [online] Available at: <https://sentinel.esa.int/web/sentinel/home> [Accessed 15 February 2021].

Appendix C.   Understanding sensors' effectiveness for bushfire prediction across the spectrum

Reduced density of vegetation increases the visible reflectance since the non-vegetated surface contributes to the reflection. This converges the visible and near IR reflectance as cover decreases (Paltridge and Barber, 1988).

Appendix Table C1: Bushfire prediction characteristics of sensor spectrum bands

Sensor-type/ description	Relevance to bushfire relevance	Pros and cons
Visible (400-700 nm)	Can detect burnt area; low levels of FMC well, but since chlorophyll absorbs solar radiation, visible reflectance of high FMC is low (Paltridge and Barber, 1988).	• Detection increases as vegetation moisture content decreases • Images can be unavailable at critical times due to cloud cover (Ban *et al.*, 2020).
Near IR (700-1,100 nm) Example satellites: Landsat 7/8, WildFireSat (proposed)	Can detect moist vegetation well (little radiation absorbed in NIR, so reflectance is high compared to visible (Paltridge and Barber, 1988)).	• Reflectance decreases as vegetation dries out and chlorophyll decreases (Paltridge and Barber, 1988).
Shortwave IR (1,100-3,000 nm) Example satellites: Landsat 7/8, Sentinel-2	Fuel assessment: Detects canopy height, density, and cover (Yebra et al., 2015)	• Limited spatial and temporal coverage (airborne lidar, Chuvieco *et al.*, 2020). • Datasets can be very large (Yebra *et al.*, 2015, pg 6) • Provides little information to distinguish between fuel types, and does not provide information on the water content of the fuel (Yebra *et al.*, 2015, pg 23) • Water absorption increases significantly at 1,400, 1,900, and 2,400 nm (NASA, 2014)

Microwave (Radar) Example satellite: RADASAT	Good for physical detection including burnt area, but is complex and difficult to interpret results across various factors. For instance, C-band is sensitive to variations in backscatter from moisture variation (forest floor, the canopy, and rain). Single channel C-band SAR can also be restricted due to variations in surface roughness and biomass confounding the results. New poliarmetric, X-, C-, and L-band SAR may be able to decompose the scattering mechanisms to improve variable extraction (Leblon et al., 2012).	• Can detect during both the day and night, and C-band is not inhibited by clouds • Availability of conditions during cloud/smoke, and at night can improve the speed and efficiency of emergency response (Ban et al., 2020). • Removal of canopy and decreased soil moisture can increase C-band backscatter, increasing burnt area detection sensitivity (Ban et al., 2020)1. • Images can be harder to interpret (Australian Government, 2020).
LiDAR (crosses visible and IR spectrum bands above) Example satellite: AEOLUS	LiDAR can record critical properties such as the amount, location, density, and arrangement of fuel components, particularly canopy height and density (Yebra et al., 2015; Chuvieco et al., 2020). Importantly, LiDAR can also be used for surface detail to complement fuel assessment. Knowledge of, for instance, creeks, rock shelves, fire trails and old tracks can assist in planning control lines for hazard reduction burning and firefighting exercises (Yebra, 2015 citing Rusell, 2007).	• Lidar's ability to detect near surface fuel (important in fire hazard and fire behaviour estimations) can be limited depending on how much vegetation is overhead (Yebra et al., 2015). • LiDAR data can be supplemented by multi-spectral data, which can provide better information on the main vegetation type (Chuvieco et al., 2020)

Appendix D.   Further information on climate, weather, and lightning for bushfire prediction

The Bureau of Meteorology (BoM), Commonwealth Scientific and Industrial Research Organization (CSIRO) and Geoscience Australia (GA) are working together to develop products for better pre-ignition fire prediction. Proactively monitoring this full suite of data contributes to ensuring preventive actions such as controlled burns in areas of excess fuel load are taken in favorable weather conditions (BoM, 2014).

The use of weather satellite data to support weather forecasting has improved the ability to forecast fire weather, and has contributed to the development of seasonal outlooks such as the quarterly *Australian Seasonal Bushfire Outlook* (Bushfire and natural hazards CRC, 2020). Fire authorities use this to make strategic decisions about resource planning and prescribed fire management.

Additionally, CSIRO collaborates with the BoM to produce national climate projections and modelling for Australia. Spark, developed by CSIRO's Data61, is one example of a modelling tool which integrates fire weather data with geographic information and fire spread models (CSIRO, 2021).

The El Niño-Southern Oscillation (ENSO) plays a critical role in influencing weather conditions in Australia, with El Niño conditions leading to reduced rainfall, rise of temperatures, and later monsoon onset. Thus, an El Niño event contributes to conditions more favorable to bush fires, but there can be cases where a weak La Nina can play a key role in promoting fire season (BOM, 2014).

Satellite data, and ground and ship observations, are integrated into numerical weather prediction models interpreted by meteorologists. The models also incorporate physical and mathematical properties of the atmosphere. Forecasts also include land-based sensor data such as winds, wind gusts, temperature, dew point, fuel load and curing (dryness).

The BoM has access to data from the Japan Meteorological Agency (JMA's) Himawari negative eight and negative nine geostationary weather satellites through an international agreement. The JMA shares the data collected by the satellites publicly, which covers the Australian region. Processed data from these satellites is used for near real-time forecasts, and for the BoM's numerical weather prediction models (BOM, nd). The project, named 'Mapping Major Bushfires

in Australia using Near Real-time Himawari-8 Satellite Imagery',
(BigDataEarth, 2020) can be very effective with for detecting bushfires
but is not focused on predicting the bushfires before they occur.

A current limitation is the use multiple different models each
representing a different weather scenario. Model extrapolation
will be improved with better quality and more frequent input of
observational data.

Bridging weather forecasts and lightning detection required for
near-term prediction of bushfires, thunderstorm activity is currently
tracked in a number of ways including use of Global Positioning
and Tracking Systems (GPATS). Satellite images can provide early
indication of developing thunderstorms and insights into the type
of thunderstorm developing (National Severe Storms Laboratory
(NSSL), 2021). Thunderstorms may be wet or dry. Dry thunderstorms
are short-lived and tend to produce little to no rain, making them a
higher risk category for bushfires *(NSSL, 2021)*.

Appendix D References

BOM, 2014. What is El Niño and what might it mean for Australia?, <http://
www.bom.gov.au/climate/updates/articles/a008-el-nino-and-australia.
shtml> [Accessed date 10 February 2021]

Bushfire and natural hazards CRC, 2020. Australian Seasonal Bushfire
Outlook: December 2020—Feb 2021. < https://www.bnhcrc.com.au/
hazardnotes/85> [Accessed 16 Feb 21]

CSIRO, 2021. Spark. <https://research.csiro.au/spark/> [Accessed 16 Feb 21]

Appendix E. The Pathway of the EarthCARE satellite LiDAR instrument and corresponding swath

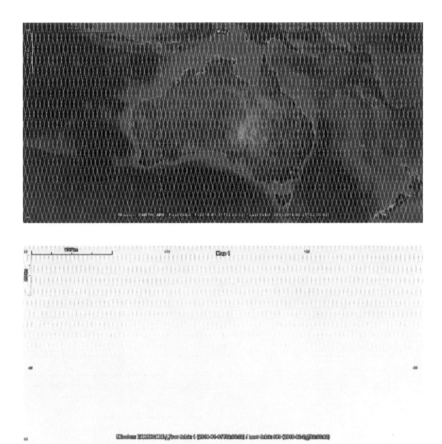

Appendix Figure E: LiDAR (top) / Multi-Spectral Imager (MSI) (bottom) instrument after one repeat cycle—25 days

(The Earth observation Swath and Orbit Visualisation tool, ESOV, 2021, https://eop-cfi.esa.int/index.php/applications/esov)

Appendix F.   Technology-enabled firefighting suit specifications

Appendix Table F1: Technology-enabled firefighting suit specifications

#	Challenge	Technology	Outcome
Personal Protection			
1	High-Stress Levels / Elevated Heart Rate	Heart-Beat Sensor using the photoplethysmography principle or blood pressure sensors	Real-time heart rate levels and trend analysis
2	High Dehydration Levels	Electrochemical dehydration sensors	Real-time hydration levels and trend analysis
3	Lost Real-Time Communications	On-board integrated computer	Core integration point of all sensor-based technologies
4	Visibility Constraints Leading to Injuries	Ultrasonic, proximity, and infra-red sensors	Enables firefighters to see into the unknown
5	Overexertion	Pedometer-like sensors	Step counts levels and trend analysis
6	Dangerous Levels of Chemical Concentrations	Electrochemical pH sensors	Ability to proactively determine dangerous chemicals
Environmental			
7	Extreme Temperatures	Temperature sensors	Real-time temperature levels and trend analysis
		Temperature-Regulators or liquid cooled suits	
8	Backdraft of Air / Winds	Ultrasonic wind sensors	Real-time wind movement levels and trend analysis
9	Significant Oxygen Deprivation	Oxygen depletion sensors	Real-time oxygen levels and trend analysis
10	Uneven / Unlevelled Terrain	Accelerometer / Gyro	Real-time terrain (x, y, z) plane and trend analysis
11	Inconsistent Humidity Levels	Humidity sensor	Real-time humidity levels and trend analysis

Advanced / Expansion Areas			
12	Inability to centrally command or control fatigued firefighters	Wireless uplink of real-time health data to a central command module at a central base	Central co-ordination centers can manage the health of the front-line workforce in real-time
13	Inability to define objects due to blurred vision	Computer vision modules	Advanced object detection due to poor line of sight
14	Inadequate data reporting or dashboards	Real-time arm-based console	Advanced real-time data feedback
15	Inability to communicate to other Firefighters or to base	Peer-to-peer 'hot-spotting' networking communication protocols	Ability to interact with Peer Firefighters in a defined radius without having to rely on established infrastructure
16	Physically weakened firefighters	Exo-skeleton developed suits with augmented robotic functionality	Super-strengthened suits to assist with movement or lifting operations
17	Inability to see from a central command station	On-board camera	Real-time video feeds to the command center
18	Requirement to continue to fight fires in oxygen-deprived areas	Oxygen supply	Access to oxygen when required
19	Constant battery discharging	Solar-shielded suits Power supply	Solar-powered recharge capabilities

Note: Red refers to Highly Complex advancements, Yellow refers to Moderately Complex advancements and Green refers to Low Complex advancements. The current design reflects the existing thermal/fire-protected Fire Fighting suits currently in use, with an additional technology enabler.

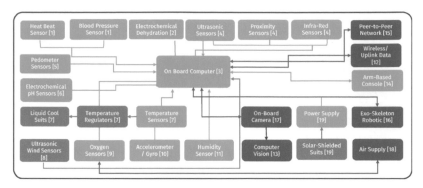

Appendix Figure F: System Integration View

Appendix G.   Australian Fire Danger Rating System (AFDRS)

The current Australian Fire Danger Rating System (FDRS) is based on two sets of fire danger indices: The Forest Fire Danger Index (FFDI) and Grass Fire Danger Index (GFDI). The indices produce a fire danger rating in one of six categories ranging from low-moderate to catastrophic. The rating is used to communicate the likelihood of a fire starting and the risks associated with controlling or suppressing it (Commonwealth of Australia, 2020; Bureau of Meteorology 2021).

The following issues have been identified with the current FDRS:

• Indices are produced using outdated scientific models from 1950s-1960s
• Indices only account for weather parameters (i.e wind speed and rainfall)
• Indices do not account for human and environmental factors (i.e fuel type and infrastructure)
• Recommended actions vary across state and territories
• Visual display varies across states and territories, as shown in Figure G below.

In 2014, the Australian Government committed to prioritize the development of a new AFDRS. A program Board was established in 2016 to oversee the development and implementation of the new system by 2023 (Commonwealth of Australia, 2020).

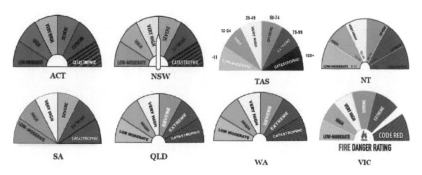

Appendix Figure G: Jurisdictional Fire Danger Rating Systems (Commonwealth of Australia, 2020)

Appendix H. The Australian Warning System

Following an extensive consultation process, a new Australian Warning System was implemented in December 2020. The changes made to the new system were based on input and feedback received from emergency services and hazard agencies as well as the community.

The system builds on existing principles to provide a nationally consistent warning system that can be applied to all hazards. The system is outlined in Appendix Figure H.

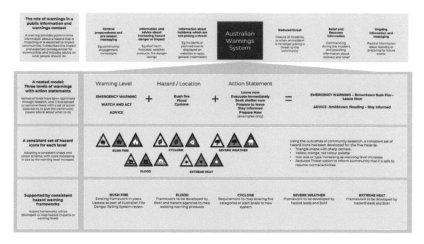

Appendix Figure H: The Australian Warning System, Obtained from the Australian Institute for Disaster Resilience, ©Commonwealth of Australia [2020]

Appendix I.   Bushfire Warning Apps

Appendix Table I1: Bushfire Warning App Specifications

	Geographic Region	Developer	Installs (Google Play)	5-Star Review Average (Google Play)	Functionality Description
Fires Near Me Australia	Australia	NSW Rural Fires Service (NSW RFS)	100,000+	4.4	Bush fire warnings and incident information using data from participating fire agencies in Australia.
Fires Near Me NSW	New South Wales	NSW Rural Fire Service	500,000+	4.4	Stay up to date on bush fires in NSW with the official NSW Rural Fire Service Fires Near Me application, providing warnings and incident information.
Alert SA	South Australia	SA Fire & Emergency Services Commission (SAFECOM)	50,000+	3.5	Provides users with timely, relevant bushfire information in South Australia. The App displays bushfire warnings and alerts, Fire Danger Ratings and Total Fire Bans, with users able to create up to 10 Watch Zones, specifying areas of interest to them.
VicEmergency	Victoria	Department of Justice & Community Safety	500,000+	3.9	VicEmergency is the official Victorian Government app for emergency warnings and information. Main Features: • Live incident map showing current warnings, planned burns and other types of incidents across Victoria.

- Simple to create a profile and use watch zones that ensure you get official warnings for your local area. Warnings and information are pushed out to a user's phone when issued by Victoria's emergency services.
- GPS integration to determine your current location and surrounding incidents.
- Emergency Warnings, Warnings, Advice and information as issued by Victoria's emergency services
- Share incidents and warnings with friends and family.
- View forecast Fire Danger Ratings on the map.
- View today's and tomorrow's Total Fire Ban status.

My Bushfire Plan	Western Australia	Department of Fire and Emergency Services (DFES)	5,000+	4.7

The My Bushfire Plan app is a bushfire preparedness tool, providing you with one place to prepare, store, share and update your bushfire plan anytime, from any device.

Appendix J.   Virtual Reality (VR)/Augmented Reality (AR) Education Tool Specifications

Space assets-supported Virtual Reality (VR) experiences for empathy development and training

Virtual Reality (VR) technology provide a total immersion into hard-to-access or dangerous places. Such experience can be accessed using a Head-Mounted-Display (HMD) including Oculus Quest and HTC Vive or be installed in an immersive CAVE (Cave Automatic Virtual Environment) room.

Gamification, storytelling and well-developed 3D interaction have the capability of developing meaningful and engaging experiences for empathy, awareness, training and simulation of bushfire hazards in the context of education, outreach and professional development of first responders (Nell Lewis, 2020). For example, the Country Fire Authority has developed *CFA Bushfire VR* learning experience to replicate a bushfire scenario in order to increase awareness of the bushfire risks (CFA, 2018). *Flaim VR* is an example of immersive training in hazardous and emergency situations developed in Australia.

The firefighters involved in the 2019 bushfires in Australia, who participated in the discussion during one of the project sessions, have expressed the lack of awareness by the population towards the bushfires. These low levels of empathy make individuals be careless about their farms and do bonfires during the fire season, which results in man-made fires. Our solution proposes to boost empathy and awareness towards bushfires for everyone in the world to feel connected with the affected people with a bush fire experience.

Several research studies (Harrington *et al*, 2016; Greitemeyer *et al*, 2011; Rowan *et al*, 2018) state that empathy can be improved through applied games. In the case of VR gamified immersive experiences, the levels of empathy are even higher (Bujić *et al*, 2020; Herrera *et al*, 2018). Furthermore, the empathy levels can last longer when there is a well-designed 3D interaction integrated; the person does not just receive information (Herrera *et al*, 2018). A VR empathy-related training is also used by the New County Police to De-Escalate situations (NBC, 2020).

Some studies even suggested empathy can be increased by using animals with facial expressions as avatars (Sierra Rativa *et al*, 2020).

Two short animations, a collaborative project by VR artists *Australian Bushfires Wildlife Awareness VR 360* (Brown, 2020) and *A Black Summer* by Sam Hoefnagels (Fresh Media, 2020), both feature iconic Australian wild animals as the main characters to develop empathy towards nature.

An Immersive Journalism is a type of journalism that creates a stronger emotional bond through a VR story (Bujić, *et al*, 2020). It has been studied to raise awareness on the 2003 SARS outbreak (Wu et al., 2020). A 360-degree video *Deadly wildfires: a devastating year for Portugal* (Euronews, 2017) is another well-developed example of immersive journalism. Satellite assets could be used in VR stories in order to provide further contextualisation.

We propose the design of an immersive VR story to increase public awareness of a bushfire risk and to better understand its behavior. The experience could be personalised through AI and GNSS location. A Google Earth VR-like or a similar platform (for example, Open Data Cube, UP42) could be used to superimpose the AI-generated fire and smoke into the 3D terrain and buildings reconstructed from the satellite data. From our program sessions with the firefighters, we've heard how the fire spreads differently depending on the slopes. Those slopes can be 3D reconstructed using AI and multi-satellite data. A personalised VR learning experience would enable anyone in the globe to experience how their home area might look and feel under the reality of a bushfire hazard.

Space assets-supported Augmented Reality (AR) applications for real-time emergency navigation and collaborative response planning

While Virtual Reality (VR) provides a complete immersion, Augmented Reality (AR) technology enable digital content to overlay the real environment. This type of information can be displayed on a smartphone, tablet or AR glasses in real-time using 4G/5G or WiFi Internet connection similar to the *Pokémon-Go* game. As described in the SNS Telecom & IT 2020 report, critical communication can be maintained through space assets. A rapid development of AR smart glasses (see Facebook Ray Ban smart glasses or Vuzix Blade Upgraded) may soon allow such wearables to become ubiquitous and a capable consumer-level alternative to MagicLeap or Hololens high-end AR headsets.

The ability of satellite-supported location-based (GNSS) AR to superimpose information in real-time can be used to identify the safest path for evacuation to assist both the affected individuals as well as first responders (Lovreglio and Kinateder, 2020). Moreover, the app could also advise on the actions and items taken before or during the evacuation process. By utilising the Internet connection, so-called social or community AR and VR, also enable shared multi-user experiences (see Spatial VR) for effective collaborative response planning (Boeing, 2016). The 3D maps generated from the satellite assets could be viewed in multiple layers connecting real-time data with historical as well as static data with dynamic. Furthermore, a live stream connection through 5G enables a new wave of social collaboration via AR/VR. As for AR heavier (not only low-poly) 3D data can be rendered in real-time, which means that the users can interact with more data and more complex and dynamic content in 3D in real-time. In addition to AR, firefighters on the ground could live stream through 5G in VR (360/real-time rendering) the situation to higher commanders in the centre. VR could also be applied to provide a plausible bushfire scenario for empathy development and education similar to an AI-supported AR application seeUV.

Satellite data-supported Artificial Intelligence to mimic bushfire environment for communication potential hazards

During an emergency crisis, knowing the behavior of wildfire is critical to emergency responders so they can orchestrate a proper response. Artificial Intelligence (AI) machine learning can reliably detect wildfires given that the dataset contains similar patterns which can help responders to plan for emergencies by conducting models of potential fire conditions before they occur (Lane, 2021). WIFIRE system is being used in California which is a hybrid of high-speed fiber optics and a fixed wireless network that links hundreds of remote weather stations throughout the county to the San Diego supercomputer hub. In real-time, WIFIRE's simulation software forecasts how a fire will spread. Through conducting models of future fire situations, the system also helps responders to plan for fires before they start (Tedjasaputra, 2020). The same system could be used to help escort victims in bushfire-affected areas to a safer location.

Appendix J References

Nell Lewis, C, 2021. Fighting wildfires with virtual reality. [online] CNN. Available at: <https://edition.cnn.com/2020/01/29/tech/virtual-reality-firefighter-training/index.html> [Accessed 15 February 2021].

Walker, A, Gard, S, Williamson, B and Bell, J, 2021. A virtual reality bushfire mitigation tool for community consultation. [online] Brisbane, Australia.: Queensland University of Technology. Available at: <http://file:///C:/Users/aljar/Downloads/a-virtual-reality-bushfire%20(1).pdf> [Accessed 9 February 2021].

Turton, B, 2020. AROC by BadVR—Revolutionizing First Responder Technology. [online] Medium. Available at: <https://medium.com/badvr/aroc-by-badvr-revolutionizing-first-responder-technology-f8d302646c57> [Accessed 11 February 2021].

Lovreglio, R and Kinateder, M, 2020. Augmented reality for pedestrian evacuation research: Promises and limitations. *Safety Science*, 128, 104750.

Morris, E, 2020. How the ABC produced the Mt Resilience Augmented Reality experience to explore a way of living with big weather and climate change. [online] Abc.net.au. Available at: <https://www.abc.net.au/news/about/backstory/2020-11-19/the-making-of-mt-resilience-ar-project/12891282> [Accessed 9 February 2021].

Ren, A, Chen Chong, C and Lou, Y, 2008. Simulation of Emergency Evacuation in Virtual Reality. *Tsinghua Science and Technology*, 651–659.

Brown, M, 2020. Australian Bushfires Wildlife Awareness VR 360 Collaboration. [online] Willowmoon Art & Design. Available at: <https://willowmoonart.com/blogs/news/australian-bushfires-wildlife-awareness-vr-360-collaboration> [Accessed 9 February 2021].

Shashkevich, A, 2018. Virtual reality can help make people more empathetic | Stanford News. [online] Stanford News. Available at: <https://news.stanford.edu/2018/10/17/virtual-reality-can-help-make-people-empathetic/> [Accessed 15 February 2021].

Piumsomboon, T, Lee, Y, Lee, G, Dey, A and Billinghurst, M, 2017. Empathic Mixed Reality: Sharing What You Feel and Interacting with What You See. 2017 International Symposium on Ubiquitous Virtual Reality (ISUVR).

Rupnik, E. Pierrot-Deseilligny, M and Delorme, A., 2018. 3D reconstruction from multi-view VHR-satellite images in MicMac. ISPRS *Journal of Photogrammetry and Remote Sensing*, 139, 201–211.

Flaimsystems.com. 2021. [online] Available at: <https://flaimsystems.com/news> [Accessed 14 February 2021].

Tuffley, D, 2019. *Virtual tools, real fires: how holograms and other tech could help outsmart bushfires*. [online] The Conversation. Available at: <https://theconversation.com/virtual-tools-real-fires-how-holograms-and-other-tech-could-help-outsmart-bushfires-126830> [Accessed 16 February 2021].

Australian emergency service magazine. 2021. CFA Embraces the Future of Virtual Firefighting Training |. [online] Available at: <https://ausemergencyservices.com.au/emergency-services/firefighters/cfa-embraces-the-future-of-virtual-firefighting-training/> [Accessed 16 February 2021].

Fresh Media, 2021. A Black Summer—VR 360 Animation. [online] YouTube. Available at: <https://www.youtube.com/watch?v=3nuLCAgiCBs> [Accessed 11 February 2021].

Harrington, B and O'Connell, M, 2016. Video games as virtual teachers: Prosocial video game use by children and adolescents from different socioeconomic groups is associated with increased empathy and prosocial behaviour. *Computers in Human Behavior*, 63, 650–658.

Greitemeyer, T and Osswald, S, 2011. Playing Prosocial Video Games Increases the Accessibility of Prosocial Thoughts. *The Journal of Social Psychology*, 151(2): 121–128.

Rowan, N, Sardina, A, Arms, T and Ashton-Forrester, C, 2018. Into Aging Simulation Improves Empthay Outcome In Allied Health Students. *Innovation in Aging*, 2 (suppl 1): 165–165.

Bujić, M, Salminen, M, Macey, J and Hamari, J, 2020. 'Empathy machine': how virtual reality affects human rights attitudes. Internet Research, 30(5): 1407–1425.

Herrera, F, Bailenson, J, Weisz, E, Ogle, E and Zaki, J, 2018. Building long-term empathy: A large-scale comparison of traditional and virtual reality perspective-taking. PLOS ONE, 13(10): p.e0204494.

Sierra Rativa, A, Postma, M and Van Zaanen, M, 2020. The Influence of Game Character Appearance on Empathy and Immersion: Virtual Non-Robotic Versus Robotic Animals. *Simulation & Gaming*, 51(5): 685–711.

Wu, H, Cai, T, Liu, Y, Luo, D and Zhang, Z, 2020. Design and development of an immersive virtual reality news application: a case study of the SARS event. *Multimedia Tools and Applications*, 80 (2): 2773–2796.

NBC10 Philadelphia. 2021. New Castle County Police Use Virtual Reality Training to Help De-escalate Situations. [online] Available at: <https://www-nbcphiladelphia-com.cdn.ampproject.org/c/s/www.nbcphiladelphia.com/news/tech/new-castle-county-police-use-virtual-reality-training-to-help-de-escalate-situations/2697112/?amp> [Accessed 16 February 2021].

Snstelecom.com. 2021. LTE & 5G for Critical Communications. [online] Available at: <https://www.snstelecom.com/lte-for-critical-communications> [Accessed 16 February 2021].

Tedjasaputra, V, 2020. Fighting Fires with Supercomputers. [online] HPCwire. Available at: <https://www.hpcwire.com/off-the-wire/fighting-fires-with-supercomputers/> [Accessed 12 February 2021].

Lane, J, 2021. FIRIS: Fire Integrated Real time Intelligence System provides real-time fire incident intelligence. [online] Csfa.net. Available at: <https://www.csfa.net/CSFA/CalFF/articles/Guest_Columns/FIRIS__ Fire_Integrated_Real_time_Intelligence_System_provides_real-time_ fire_incident_intelligence.aspx> [Accessed 12 February 2021].

Boeing, 2016. Boeing drones and Microsoft holograms to fight wildfires. [online] Available at: <https://www.commercialuavnews.com/public-safety/boeing-drones-microsoft-holograms-fight-wildfires> [Accessed 13 February 2021].

Appendix K.   Emergency Satellite Communication System Specifications

Space cell towers: Direct communication system between smartphones and satellites

This strategy aligns with the space strategic planning of the Australian Government as stated in the 'Communications Technologies and Services Roadmap 2021-2030', 13.

During emergency situations is vital that the country's communication system works properly. with this, they can broadcast information to the general public, emergency services can orchestrate a better response, and the population can keep track of their relatives. All this system is based upon different pieces of infrastructure. An important component of this network are the cell towers responsible for the routing of our everyday communications.

Bushfires are so powerful that they can destroy the cell towers. When they get tear down by wildfires, we obtain a blackout in our communication system, and thus, the government agencies cannot communicate with the general public, emergency services cannot make a proper response, and the public cannot keep track of their relatives. All this results in general fear, loss of time, loss of lives, and loss of wildlife. During the 2019–2020 Australian bushfires the state of emergency lasted weeks. There exists a need for a resilient communication solution that will withstand continuous fire (supported by the Royal Commission, 240). In addition, even if we have our cell tower at full power, the signal coverage, as we experience every day, is not enough to guarantee a complete connection everywhere at any time. In an emergency situation, it is of paramount importance to have a direct and reliable link with emergency teams.

Australia is in a unique position, where mobile phone connection only covers a small section (twenty-seven per cent) of the country. Nevertheless, Australia is not alone in this problem, US Public Utilities Commission reported that 2017 Northern California wildfires left 85,000 wireless customers without service. This statistic included 11–15 Public Safety Answering points, which are critical point of call during emergency.

Improvement of signal coverage is a mission of several companies in Australia, including Fleet, Inovor and Myriota. Fleet is providing solutions to connect individuals living in rural and remote areas of Australia with reliable internet connection through their fleet of nano-satellites in Low Earth Orbit (LEO). Inovor's mission sees satellite data connected with machine learning technology to provide their users with custom, informative space data. The problem of several of this solution is that they require the end user to buy a satellite phone terminal. Here we propose a solution of emergency and daily based space communication system via a direct link between non-modified smartphones and satellites, which work as space sell towers. This concept has been shown to work by Lynk [https://lynk.world/], they have successfully broadcasted an emergency SMS directly from the satellite towers to a non-modified smartphone. The government can also use this to broadcast information to a custom app like WhatsApp, so users can interact easily with the emergency services and their relatives.

Lynk: connecting unmodified phones to satellites

An aerospace startup called Lynk plans to launch thousands of satellites 'cell towers' in order to connect common phones to them and solve the coverage issues that we experience every day. They have already tested their system from the ISS and they were able to send the first text message to a common Android smartphone from space. Their goals are global cellphone connectivity from orbit with a wide range of applications.

One of the strengths of their project is that customers won't need to provide extra hardware. It will be compatible with all the standard cellphones that we are using nowadays.

'*No one ever in human history has used a satellite to send a message directly to an unmodified mobile phone on the ground,*' Charles Miller, co-founder and CEO of Lynk, tells The Verge. Up to now, if you want to use a satellite phone, you have to possess a special phone or equipment which are not compatible with our common smartphones. The company says that they developed a software able to let our phones connect to satellites 'tricking' the signal as if it was emitted from a standard ground-based cell tower. The handover: change in phone signal from tower to satellite, occurs automatically whenever the phone cannot reach a land base cell tower. The phone will connect directly to the satellite signal, and will provide SMS connectivity straight to the smartphone.

To test their technology, Lynk launched its third payload to the International Space Station in December 2019 aboard a SpaceX Falcon 9 rocket. They attached the payload outside of the ISS on a Cygnus spacecraft. It detached from the space station and, staying in orbit, allowed the test. At the end of February, the first text from space was received by an Android phone in Falkland Islands. It is a milestone.

It is a crucial technology demonstration that allows the company to move forward with their pro ject. Their plan is to start to put in orbit mini-satellites of about fifty-five pounds (twenty-five kilograms) in LEO. Their initial test has proven viable to broadcast emergency services messages to non-modified smartphones directly through a satellite. The data speed that they achieved up to now is comparable to a 2G, but with thousands of satellites they eventually will be able to achieve 4G coverage or more.

By connecting more people around the world, the company is helping to overcome the digital divide, which disproportionately affects developing nations and communities with low socioeconomic status. Connected communities can innovate faster and participate in the digital economy, which drives economic growth.

Appendix K References

1. Australia mobile coverage: https://www.clientsat.com.au/mobile-phone-coverage-in-australia/
2. https://lynk.world/
3. https://www.google.it/amp/s/www.theverge.com/platform/amp/2020/3/18/21184126/lynk-mega-constellation-text-message-android-smartphone-cell-towers-space

4. https://www.nasa.gov/sites/default/files/atoms/files/space_portal_cst_lynk_global.pdf (GOOD LINK FOR GRAPHICS)
5. https://www.cpuc.ca.gov/uploadedFiles/CPUC_Public_Website/Conte nt/Safety/Telco% 20Fire.pdf
6. https://ecfsapi.fcc.gov/file/1053130424752/EAS-1.-NBNCBC-Telecommunications-Outage-Report-2017-Firestorm.pdf

Appendix L.   The Disaster Charter

As discussed in the report, the technologies in predicting and mitigating bushfires pertain mostly to the use of remote sensing for the prediction and management of disasters. It is important to note that there are two (2) sets of international rules which may apply in this type of space science technology—The Principles Relating to Remote Sensing the Earth from Outer Space, UN Doc A/RES/41/65/ (1986) also known as Remote Sensing Principles (hereinafter referred to as 'Remote Sensing Principles') and the International Charter on Cooperation to Achieve the Coordinated Use of Space Facilities in the Event of Natural or Technological Disasters also known as Disaster Charter (hereinafter referred to as 'Disaster Charter'). The Remote Sensing Principles apply to all non-security related remote sensing of Earth from space (Monteshar, 2016). Meanwhile, the Disaster Charter covers not only remote sensing but also other space-based services of use in disaster situations (Monteshar, 2016). For purposes of this paper, further discussion shall be made on how the Disaster Charter may serve as an additional mechanism of early warning system and of predicting disasters, for consideration of policy makers, space agencies, and the government in general.

As a background, the Disaster Charter was initiated by ESA and the French space agency, CNES and formally began operating in November 2000 (Monteshar, 2016). The establishment of the Disaster Charter stemmed from the appreciation of the usefulness and availability of Earth observation information in early warnings and managing the impact of natural or technological disasters (Monteshar, 2016). It is also important to note that Article 2 of the Disaster Charter defines the term 'natural or technological disaster' as a situation of great distress involving loss of human life or large-scale damage to property, caused by a natural phenomenon, such as a cyclone, tornado, earthquake, volcanic eruption, flood or forest fire, or by a technological accident, such as pollution by hydrocarbons, toxic or radioactive substances.

Unlike the Remote Sensing Principles, the members of the Disaster Charter are not States. Instead, these are agencies, or national or international space system operators with access to space facilities (Article VI, Disaster Charter) These space facilities are not limited to remote sensing but may also refer to other space systems for observation, meteorology, positioning, telecommunications and TV broadcasting or elements thereof such as on-board instruments, terminals, beacons, receivers, VSATs and archives (Article I, Disaster Charter).

Having 17 Charter Members (The International Charter Space and Major Disasters, 2020) in the Disaster Charter as of writing, below are its Purposes pursuant to Article II:

1. Supply during periods of crisis to States and communities whose population, activities or property are exposed to an imminent risk, or are already victims, of natural or technological disasters, data providing a basis for critical information for the anticipation and management of potential crises; and
2. Participation, by means of this data and of the information and services resulting from the exploitation of space facilities, in the organization of emergency assistance or reconstruction and subsequent operations.

It is important to note that *crisis* in the above enumeration refers to the period immediately before, during or immediately after a natural or technological disaster, in the course of which warning, emergency or rescue operations take place (Article I, Disaster Charter).

Building on the above-mentioned purposes, the Disaster Charter mandates that there must be a clear description of space assets and extent of the coverage that they may provide so that it may be of effective assistance to countries in crisis. An important operation of this Charter in mitigating bush fires would be Scenario-Writing as provided in Article 4.2. Charter Parties are encouraged to keep abreast of new methods being developed in applied research for warning of, anticipating and managing disasters (Article 4.2, Disaster Charter). It is also worth mentioning that the Disaster Charter provides for waiver of the liability of satellite operators who provide data under said Charter (Montershar, 2016).

The Charter has supported forty-nine wildfire cases since its initiation in 2000 by providing imagery from multiple satellites to assess the scope and analyze damages from wildfires (The International Charter Space and Major Disasters, 2020). As of August 2020, there have been 670 activations of the Charter, and approximately eight per cent of those activations have been for wildfires. forty-five per cent of the wildfire activations requested were in South America, Europe and North America (The International Charter Space and Major Disasters, 2020).

On 13 November 2019 (The International Charter Space and Major Disasters, 2019), the Geoscience Australia on behalf of Emergency Management Australia Crisis Coordination Centre (CCC) and New South Wales Rural Fire Service (NSWRFS) has activated the Disaster Charter covering bushfires which have spread across two states of Australia burning over a million hectares of land. As soon as an activation was made, Charter member agencies and partners provided 2,077 products collected from twenty satellites (The International Charter Space and Major Disasters, 2020). This marked the largest number of satellite products provided for any activation in the Charter's history at the time (The International Charter Space and Major Disasters, 2020).

Considering the above, it is evident that the Disaster Charter has the capacity to provide various satellite images as valuable information in disaster risk management. As such, policy makers should consider the regulation of these kinds of technologies so as to activate its benefits and at the same time encourage cooperation and sharing of information.

Appendix L References

Mosteshar, S. *Regulation of remote sensing by satellites*, in Routledge Handbook of Space Law, November 2016, p.144

The International Charter Space and Major Disasters. (10 February 2021). *Membership History*. https://disasterscharter.org/web/guest/history

The International Charter Space and Major Disasters. (20 August 2020). *Fire monitoring through the Charter*. https://disasterscharter.org/web/guest/-/fire-monitoring-through-the-charter

The International Charter Space and Major Disasters. (13 November 2019). *Fire in Australia.* https://disasterscharter.org/web/guest/activations/-/article/fire-in-australia-activation-631-

Appendix M.   Examples of collaborative space innovation

Some examples under development include:

- Planet.com is developing a space-based Earth observation map to enable fire management services to better estimate fire risk throughout California (Peters, 2020).
- Danish Startup, Robotto, in collaboration with Danish Emergency Management Agency is developing an autonomous drone which uses AI to provide a faster and more accurate way to measure fires (Foverskov, 2020)
- Howe and How Technologies are developing a firefighting robot named Thermite, that's able to withstand extreme fire conditions (Orange, 2017)
- Virtual Reality (VR) experiences for empathy development and training
- Innovative solutions to promote bushfire preparedness and support early evacuation behavior (Immersive VR and AR technology)
- Augmented Reality (AR) applications for real-time emergency navigation and collaborative response planning
- 5G connection for communicating complex space-born data through immersive real-time experiences

CPSIA information can be obtained
at www.ICGtesting.com
Printed in the USA
BVHW021736291021
620151BV00001B/21